Steel Structures – Analysis and Design for Vibrations and Earthquakes

Based on Eurocode 3 and Eurocode 8

Karuna Moy Ghosh

CEng, FIStructE (London), MIE Civil Engg. (India), Chartered Civil and Structural Engineer,
Formerly, Chief Structural Engineer, Kaiser Engineers and Constructors Inc.

Whittles Publishing
CRC Press
Taylor & Francis Group

Published by
Whittles Publishing,
Dunbeath,
Caithness K W6 6EG,
Scotland, UK

www.whittlespublishing.com

Distributed in North America by
CRC Prcss LLC,
Taylor and Francis Group,
6000 Broken Sound Parkway NW, Suite 300,
Boca Raton. FL 33487, USA

ISBN 978-184995-035-0
USA ISBN 978-1-4398-8347-1

Front cover photograph by kind permission of Cliff Chamberlain, www.3c-communicalions.co.uk

Printed and bound by Bell & Bain Ltd., Glasgow

Contents

Preface

In this book the author addresses the behaviour of a structure under the action of earthquakes and the dynamic effects of unbalanced rotating masses of machines. These are the two most common complex vibration problems. Seismic motions of the ground cause a structure to vibrate. Similarly, an unbalanced rotating mass of a machine generates vibrations and transmits these to the supporting structure.

So, before we proceed to analyse and design the structure, it will be the chief objective to evaluate the design parameters so that a safe, sustainable structure can be designed. In order to evaluate the design parameters, we should have a clear understanding of the concepts of seismology and the fundamentals of moving periodic or impact loads (dynamic loads) acting on a machine supporting structure.

The author has selected two case studies, namely:

- A steel-framed storage building subjected to earthquake forces.
- A steel-framed structure subjected to machine vibrations.

Before the analysis and structural design, the design parameters are described and established. The analysis and design are then carried out based on the following latest codes of practices:

- Eurocode 3 (EC3): Design of steel structures.
- Eurocode 8 (EC8): Design of structures for earthquake resistance.

This book has been written with the structural calculations presented in a simple and clear way, with step-by-step procedures stating the design philosophy, functional aspects, selection of construction material and method of construction. The relevant clauses of the above codes of practice are clarified, and sufficient design sketches, tables and references are included where necessary.

It is believed that this book will be useful as a practical design guide and reference text for advanced university students, newly qualified university graduates working in a consulting design office and candidates preparing for professional examinations.

Author's note

To facilitate ease of calculation and compliance with Eurocode 3 and Eurocode 8, equation numbers provided in the text are those used in the codes.

PART I

ANALYSIS AND DESIGN OF A STEEL-FRAMED STORAGE BUILDING SUBJECTED TO EARTHQUAKE FORCES

This part of the book addresses the behaviour and effect of earthquake forces on structures founded on soil. Earthquakes are natural phenomena that generate a dynamic ground wave motion resulting in massive damage to and destruction of buildings, bridges, dams and human life. This can happen at considerable distances from the source region (epicentre), due to the vibration caused by the elastic waves in the land or to the effects of water waves (tsunami) in the ocean, depending on the severity of the ground motion and magnitude of the forces.

CHAPTER 1

Evaluation of Seismic Design Parameters

Our main objective first will be to evaluate the seismic design parameters that will enable us to design a safe and sustainable structure against the seismic forces of any magnitude resulting from an earthquake. To evaluate the seismic design parameters we should have a clear understanding of the following aspects of seismology.

1.1 Causes of earthquakes

There are two main hypotheses for the deformation of the Earth's surface (such as mountain building) due to earthquakes:

(1) Radial contraction or expansion of the whole Earth.
(2) Slow convective motion of the material within the Earth's mantle.

It may be possible for either or both of these processes to occur at any geological period of history. Neither hypothesis, however, has been developed fully to the point where it is possible to predict all the present pattern of stresses and temperature in the Earth's crust and upper mantle.

Taking into account *regional strain* (*deformation*) patterns indicates that shallow earthquakes provide the information of *global tectonic processes* with the result of a fault – a fracture or fracture zone along which the two sides have been displaced relative to one another, parallel to the fracture. Another cause of earthquakes has been studied by Matuzawa (1964) in his work based on *thermodynamical theory*. The globe is considered as a heat engine which, by doing work, produces earthquakes.

1.2 Types of earthquake

There are two main types of earthquake, namely:

(1) **Type 1:** Earthquakes of shallow focus and moderate to large energy release are accompanied by pronounced ground deformation at the surface. This deformation often takes the form of a fault rupture, surface uplift or subsidence. The earthquake that occurred in Alaska in 1964 with a Richter scale magnitude of 8.4 was caused by the deformation of a surface fault, with the length of the surface rupture of approximately 62 km and a displacement of the main fault of 6.7 m.
(2) **Type 2:** Earthquakes that release a relatively moderate to large amount of seismic energy may occur both under continents and under oceans. The causes of most long waves (*tsunamis*) are large submarine earthquakes due mainly to submerged tectonic displacements of the ocean floor.

> In 2004 in south-east Asia and in 2011 in Japan, with ground acceleration of 8.9 in Richter scale, immense devastation was caused by 10-12m high Tsunami waves after large submarine earthquakes initiated by enormous submerged tectonic displacements. The epicentre was about 80 miles from the city of Sandai on the north-east coast of Japan, near Tokyo.

Earthquake *foci* (*epicentres*) vary in depth from near ground surface to depths of about 700 km. The *frequency distribution* of earthquakes as a function of focal depth is neither uniform with depth nor uniform with geographical region.

1.3 Intensity of an earthquake

If any instrumental recordings of the ground motion are not available, seismologists describe the severity (intensity) of ground shaking by assigning *modified Mercalli (MM) intensity numbers*. The MM scale ranges from I (ground motion not felt by anyone) to XII (total damage), as shown in Table 1.1.

1.4 Magnitude of an earthquake

A severe high magnitude of earthquake involves a stress failure (slip) over a large fault area, a large release of strain energy in the form of seismic waves and a large area being subjected

Table 1.1. Modified Mercalli scale corresponding to severity of ground shaking

Modified Mercalli intensity scale	Description
I	Detected only by sensitive instrument
II	Felt by a few persons +at rest, especially on upper floors; delicate suspended objects may swing
III	Felt noticeably indoors, but not always recognised as an earthquake; standing persons/objects rock slightly, vibrations like from passing trains
IV	Felt indoors by many, outdoors by a few; at night may awaken some people; dishes, windows, doors disturbed; cars rock noticeably
V	Felt by most people; some breakage of dishes, windows and plaster; disturbance of tall objects
VI	Felt by all; many are frightened and run outdoors; falling plaster and ceilings; minor damage
VII	Everybody runs outdoors; damage to buildings varies, depending on the quality of construction; noticed by drivers of cars
VIII	Panel walls thrown out of frames; walls, monuments, and chimneys fall down, sand and mud ejected; drivers of cars disturbed
IX	Buildings shifted off foundations, cracked, thrown out of plumb; ground cracked; underground pipes broken
X	Most masonry and frame structures destroyed; ground cracked; railway tracks bent; landslides
XI	New structures remain standing; bridges destroyed; fissures in the ground; pipes broken; landslides; railway tracks bent
XII	Total damage; waves seen on the ground surface; lines of sight and levels distorted; objects thrown up into the air

to strong ground shaking. It is important for engineering purposes to be able to describe in a quantitative way the size of the earthquake.

In 1935, C. F. Richter (1935) of California Institute of Technology expressed the magnitude of an earthquake for shallow shocks in the following form:

$$M = \log_{10} \times A/A_0$$

where

M = magnitude of the earthquake – conveniently called the *Richter scale*,
A = maximum amplitude recorded by a Wood-Anderson seismograph at a distance of 100 km from the centre of the disturbance,
A_0 = amplitude of 1/100 of a millimetre.

In practice, the recordings should be made at large enough distances in comparison to the slipped fault zone. The results for the value of M can be obtained from a number of recordings from various seismological stations.

For earthquakes of Richter scale magnitude less than 5.0, the ground motion is unlikely to be damaging because of its very short duration and moderate acceleration. Earthquakes of magnitude 5.0 and greater generate severe ground motions resulting in potential damage to structures.

Thus, the use of the Richter scale is a convenient way of classifying earthquakes according to their magnitude. But, due to nonuniformity of the Earth's crust, different fault orientations, etc., the Richter scale is not a precise measure of the size of an earthquake.

1.5 Fault rupture parameters

A *fault* is defined as a fracture or fracture zone along which the two sides have been displaced relative to one another, parallel to the fracture. The displacement may range from a few centimetres to many kilometres.

Faults and faulting are of great importance to structural engineers for the following reasons:

- They may severely damage and ultimately destroy structures, due to shearing, compression, extension and rotation, caused by tilting or bending.
- Earthquakes can occur along them.
- Faulting in the past may have severely affected the physical properties of foundation soils by decreasing their shear strength, changing their permeability, or bringing together rock units with very different physical properties.

Surface manifestations of faulting and closely related processes include sudden rupture and displacement, creep, warping, tilting, and gross changes in land level. Of most significance in engineering structures is the sudden rupture and displacement of the Earth's surface, which occurs with normal, reverse, strike slip or oblique slip faulting.

The history of surface faulting in the continental U.S.A. and most of the other continents has already been recorded. The length of surface ruptures is the distance between the ends of continuous or nearly continuous breaks that formed at the surface in the listed earthquakes given by M. G. Bonilla (1966). The recorded surface fault displacement accompanying earthquakes varies from 1.5 cm of strike slip in the California earthquake of 1966, to 128 cm of vertical displacement in the Yakutat Bay, Alaska, earthquake of 1899. The length of subsurface faulting that occurred in a 1964 Alaskan earthquake was approximately 590 km, as recorded by Savage and Hastie (1966).

The width of the rupturing surface zone varies from a few centimetres to 1.5 metres or more. Fault ruptures may consist of a single narrow main break, but often they are more complicated and accompanied by subsidiary breaks. In addition to shearing displacements, surface faulting is commonly accompanied by extension or compression approximately perpendicular to the fault.

1.6 Ground motion

The *ground motion* may be defined as the shaking of the ground surface during an earthquake and is generated by the passage of stress waves, as explained by Richter (1958). These seismic waves originate from a region of the Earth's crust where a stress failure resulted from a sudden change in the equilibrium stress state. The size of earthquakes and the frequency of occurrence depend on the state of stress in the Earth's crust.

The frequency of occurrence of earthquakes in some regions results in the release of a relatively large quantity of strain energy over the years, and this suggests that there is a process that is putting strain energy into the Earth's crust at approximately the same average rate as energy is being released by earthquakes. It has also been suggested that a viscous flow in the Earth's interior is the active process responsible for straining the Earth's crust (Allen et al., 1965; Ryall et al., 1966).

1.7 Geotechnical data

1.7.1 Dynamic effects on the soil

Earthquake tremors and vibrations significantly affect the behaviour, resistance and bearing capacity of soil which supports foundations and superstructures. Results of dynamic soil investigations show that the transmission of vibrations (emanating from the seismic ground accelerations) to the soil considerably changes its behaviour and elastic properties. These changes of the soil depend on the seismic forces of periodically alternating intensity applied at the ground surface, more precisely the frequency and magnitude of vibration transmitted to the soil.

1.7.2 The natural frequency of soil

The *natural frequency of soil* varies for different types of soil. Dense soils have high natural frequencies and loose soils have correspondingly lower values. The vibrations of soil are damped by internal friction. This is the work necessary to overcome the resistance between the soil particles tending to pack together, or, in other words, the work necessary to destroy the cohesion between the soil particles – the shearing resistance of the soil. The work done results in permanent deformation.

It has also been observed that in heavy soil the velocity of propagation of periodical vibrations is higher than in light, and loose, sand. The higher frequency and amplitude of vibration reduces the permissible soil-bearing stress (bearing capacity), as shown in Table 1.2. The velocity of propagation of waves yields useful information regarding the relative bearing capacity of soil. The greater the velocity of propagation of waves, the greater the bearing capacity of soil.

Table 1.2. Relationship between velocity of wave propagation and soil stress

Type of soil	Velocity of wave propagation v (m/s)	Permissible soil stress (kN/m^2)
3 m of peat on sand	80	0
Silty sand	110	100
Wet sand	140	200
Dry sand	160	200
Silty sand on marl	170	250
Coarse sand	180	250
Alluvial marl	190	300
Homogeneous sand	220	400
Gravel under 4 m sand	330	450
Dense coarse gravel	420	450

Table 1.3. Geotechnical soil parameters

Soil type	Depth below GL (m)	γ_b (kN/m^3)	ϕ' (degrees)	δ (degrees)	N	N_q
Loose to medium clay sand	0.0	18	25	20	10	10
Medium to dense clay sand	2.0	21	30	22.5	20	20
Dense to very dense sand	9.0	21	35	36.3	30	40
Silty decomposed granite, medium to coarse grained	12.0	21	40	30	40	60

1.7.3 Dynamic soil investigations

In order to perform a dynamic soil investigation, a basic soil exploration at the site should be carried out first. Geotechnical soil parameters derived from dynamic laboratory experiments for soil types at different depths below ground level (GL) are given in Table 1.3.

1.7.4 Ground conditions

Appropriate investigations shall be carried out in order to identify the ground conditions.

- The construction site and the nature of the supporting ground should normally be free from risks of ground rupture, slope instability and permanent settlements caused by liquefaction in the event of an earthquake. The possibility of occurrence of such phenomena shall be investigated.
- Depending on the importance class of the structure and the particular condition of the project, ground investigations and/or geological studies should be performed to determine the seismic action.

1.7.5 Calculation of the ultimate bearing capacity of soil

The dynamic effects on the soil reduce the friction considerably. Owing to the energy of vibration, the coefficient of internal friction decreases, as does, obviously, the bearing capacity of soil. The dynamic soil experiments have shown that the shearing strength of soil decreases as much as 25% to 30%, depending on the acceleration of the vibration. Friction decreases as the acceleration of vibration increases.

Considering the properties of soil at a depth of 3.0 m that supports the foundation, the coefficient of internal friction (ϕ') shall be reduced. Assume the reduction factor to be 30%. Therefore the reduced value of $\phi' = 30° \times 0.3 = 21°$. With this reduced value of $\phi' = 21°$, we obtain the value bearing capacity factors N_c, N_q and N_y from Terzaghi's bearing capacity diagram,

$$N_c = 18; N_q = 8; N_y = 4.5.$$

Terzaghi suggested for square footing the following expression:

ultimate bearing capacity $q_u = 1.3 \times C_u \times N_c + \gamma_b \times z \times N_q + 0.4 \times \gamma_b \times B \times N_y$

where

C_u = undisturbed, undrained shear strength of cohesive soil = 0.0 (assumed),
z = depth of foundation in m from ground level,
γ_b = bulk density of soil in kN/m³ = 21 kN/m³,
B = width of foundation in m = 4.0 m (assumed),
N_c, N_q, N_y = bearing capacity factors,
ground water level = 3.5 m from ground level (assumed).

Therefore

$$q_u = 1.3 \times 0 + 21 \times 3 \times 8 + 0.4 \times 21 \times 4 \times 4.5 = 655 \text{ kN/m}^2.$$

Assuming a safety factor of 2.5, the safe bearing capacity of the soil is

$$q = \frac{655}{2.5} = 262 \text{ kN/m}^2,$$

whereas without any dynamic considerations the safe bearing capacity would have been 500 kN/m² with $\phi' = 30°$.

1.8 Response spectrum

The *elastic response spectrum* may be defined as the earthquake motion at a given point on the surface, represented by an elastic acceleration response spectrum. The response spectrum is a convenient way of describing a particular earthquake ground motion by plotting the maximum response velocity, response acceleration or response displacement of an oscillator, and it should be noted that this is an instantaneous value that may not be approached again during an earthquake.

For example, the time–history of motion of an oscillator having a period $T = 1.0$ sec may show a maximum acceleration of 25% g, which is the response spectrum value, whereas the second and third accelerations might be 20% g and 15% g, where g is the acceleration of gravity.

- The shape of the elastic response is taken according to the following requirements:
  - (a) No-collapse requirement (ultimate-limit-state design seismic action) and for the damage-limitation requirement with adequate degree of reliability.
  - (b) Damage-limitation requirement.

- The horizontal seismic action is described by two orthogonal components assumed to be independent and represented by the same response spectrum.
- Time–history representations of the earthquake motion may be used.
- For important structures when $\gamma_I > 1.0$, topographic amplification effects should be considered (where γ_I the importance factor which relates to the consequences of a structural failure, see Chapter 4).

The shape of the elastic response spectrum can be seen in Fig. 3.1 of EC 8: Part 1 (Eurocode, 2004).

1.9 Design spectrum

Fundamentally, vibration is a problem in the behaviour of a structure during an earthquake. The seismic motions of the ground cause the structure to vibrate, and the amplitude and distribution of dynamic deformation, and the duration of the seismic motions, are of concern to the engineer.

The primary objective of earthquake code requirements is that the structure should not be a hazard to life in the event of strong ground shaking. During moderate ground shaking that has a significant probability of occurrence during the life of the structure, the vibrations may be in the elastic range with no damaging amplitudes. However, during strong ground shaking, the structural members may undergo plastic strains and some cracking may develop. Calculations should be made for the earthquake-induced vibrations of the structure, and these will show the general nature and amplitude of the deformations that can be expected during earthquake ground shaking.

The basic earthquake design criteria shall be based on the following considerations:

- The probability of occurrence of strong ground shaking.
- The characteristics of the ground motion.
- The nature of the structural deformations.
- The behaviour of building materials when subjected to transient oscillatory strains.
- The nature of the building damage that might be sustained.
- The cost of repairing damage as compared to the cost of providing additional earthquake resistance.
- A specification of the desired strength of structures – there will be a uniform factor of safety for different structures as well as for different parts of the same structure.

The above design criteria shall be established by means of a design spectrum for elastic analysis. A design spectrum analysis shall be based on the following criteria:

- The capacity of structural systems to resist seismic actions in the non-linear range generally permits their design for resistance to seismic forces smaller than those corresponding to a linear-elastic response.
- To avoid complex inelastic structural analysis in design, the capacity of the structure to dissipate energy, through mainly ductile behaviour of its elements and/or other mechanisms, is taken into account by carrying out an elastic analysis based on a response spectrum that is reduced with respect to the elastic one. This is defined as a *design spectrum*, and the reduction is achieved by introducing a behaviour factor q.

- The behaviour factor q is an approximation of the ratio of the seismic forces that the structure would experience if its response was completely elastic with 5% viscous damping, to the seismic forces that may be used in the design with a conventional elastic analysis model, still ensuring a satisfactory response of the structure.

References

Allen, C.R., Amand, P. St, Richter, C.F and Nordquist, J.M., 1965. Relationship between Seismicity and Geologic Structure in the Southern California region, *Bull. Seism. Soc. Am*, 55(4), 753–797.

Bonilla, M.G., 1966. Deformation of Railroad Trucks by Slippage on the Hayward Fault in the Niles District of Fremont, California, *Bull. Seism. Soc. Am*, 56, 281.

Eurocode, 2004. BS EN 1998-1: 2004, Eurocode 8. Design of Structures for Earthquake Resistance.

Matuzawa, T., 1964. *Study of Earthquakes*, Uno Shoten, Tokyo.

Richter, C., 1958. *Elementary Seismology*, Freeman, San Francisco.

Ryall, A.D., Slemmons, D.B and Gedney, L.D., 1966. Seismicity, Tectonism, and Surface Faulting in the Western United States During Historic Time, *Bull. Seism. Soc. Am*, 56(5), 1105–1135.

Savage, J.C. and Hastie L.M., 1966. Surface Deformation Associated with Dip-slip Faulting, *J. Geophys. Res.*, 71(20), 4897–4904.

Terzaghi, Karl and Peck, Ralph B., 1962. *Soil Mechanics in Engineering Practice*; Wiley International, USA.

CHAPTER 2

Specifications and Rules for Steel-Framed Building Design

2.1 Materials

Referring to Clause 6.2 of EC 8 (Eurocode, 2004):

(1) Structural steel shall conform to standards referred to EC 3, Part 1-1 "Design of steel structures" (Eurocode, 2005).

(2) The distribution of material properties, such as yield strength and toughness, in the structure shall be such that dissipative zones form where they are intended to in the design.
(*Note:* Dissipative zones are expected to yield before other zones leave the elastic range during an earthquake.)

(3) The requirement of (2) may be satisfied if the yield strength of the steel of dissipative zones and the design of the structure conform to one of the following conditions:

- The actual maximum yield strength $f_{y,\,max}$ of the steel of dissipative zones satisfies the following expression

  $$fy,_{max} \leq 1.1 \times \gamma_{ov} \times f_y$$

  where

  γ_{ov} = overstrength factor used in the design,
  f_y = nominal yield strength specified for the steel grade.

  The recommended value of $\gamma_{ov} = 1.25$. So, for steel grade S 275 and with $\gamma_{ov} = 1.25$,

  $$1.1 \times \gamma_{ov} \times 275 = 1.1 \times 1.25 \times 275 = 378 \text{ N/mm}^2.$$

  Thus, the condition is satisfied if the maximum value

  $$f_{y,max} \leq 378 \text{ N/mm}^2.$$

- In the design of the structure, just one grade of steel shall be used with nominal yield strength. The upper value $f_{y,max}$ for dissipative zones and nominal value f_y shall be used.

In our case, we shall satisfy both the conditions above in the design.

2.2 Structural types and behaviour factors

Referring to Clause 6.3 of EC 8 ("Structural types and behaviour factors"), in steel-framed buildings there are two main types of structure used:

(1) **Moment-resisting frames:** In moment-resisting frames, the horizontal forces are mainly resisted by members acting in an essentially flexural manner.
(2) **Braced frames:** In braced frames, the horizontal forces are mainly resisted by members subjected to axial forces.

2.2.1 Moment-resisting frames

In moment-resisting frames, depending on the selection of ductility class, a behaviour factor is adopted. For a high-ductility class (DCH) structure, a high behaviour factor $q = 4$ or greater is used, whereas in a medium-ductility class (DCM) the upper limit of the behaviour factor is taken to be $q = 4$.

For regular structures in areas of low seismicity, a value of $q = 1.5$ is adopted without applying dissipative methods. In areas of high seismicity, a higher behaviour factor coupled with sufficient ductility within dissipative zones should be adopted.

There is a direct relationship between local buckling and rotational ductility. The dissipative zones should satisfy cross-section classification depending on the value of q. In moment-resisting frames, the dissipative zones should mainly be located in plastic hinges at the beam ends or the beam-column joints, so that the energy is dissipated by means of cyclic bending. For columns, the dissipative zones may be located in plastic hinges at the base and in the top storey.

2.2.2 Braced frames

In the case of typical braced frames, the ductility class DCH or DCM is applicable in which the behaviour factor used for DCM satisfies $1.5 < q \leq 4$ and for DCH $q > 4$, depending on the cross-sections classification in dissipative zones.

The dissipative zones are located mainly in tension diagonals, neglecting the compression diagonals. For V bracings, the horizontal forces can be resisted by taking into consideration both tension and compression diagonals. The intersection point of these diagonals lies on a horizontal member which shall be continuous.

Table 6.2 in Chapter 6 shows the upper limit of behaviour factors for moment-resisting and concentrically braced frames.

2.3 Design for moment-resisting frames

2.3.1 Beams (referring to Clause 6.6.2 of EC 8)

Beams should be verified as having sufficient resistance against lateral and torsional buckling in accordance with EC 3, Part 1-1 (Eurocode, 2005), assuming the formation of a plastic hinge at the end of the beam. The beam end that should be considered is the most stressed end in the seismic design situation.

- For plastic hinges in the beam it should be verified that the full plastic moment resistance and rotation capacity are not decreased by the compression and shear forces. For sections belonging to cross-sectional classes 1 and 2, the following inequalities should be verified at the location where the formation of hinges is expected. See EC 8 (Eurocode, 2004):

$$M_{Ed}/M_{pl,Rd} \leq 1.0 \quad (6.2)$$
$$N_{Ed}/N_{pl,Rd} \leq 0.15 \quad (6.3)$$
$$V_{Ed}/V_{pl,Rd} \leq 0.5 \quad (6.4)$$

where

M_{Ed} = design bending moment,
N_{Ed} = design axial force,
V_{Ed} = design shear,
$N_{pl,\,Rd}$, $M_{pl,\,Rd}$, $V_{pl,\,Rd}$ = design resistances in accordance with EC 3,

and

$$V_{Ed} = V_{Ed,G} + V_{Ed,M}, \quad (6.5)$$

$V_{Ed,G}$ = design value of the shear force due to the non-elastic action,
$V_{Ed,M}$ = design value of the shear force due to the application of the plastic moments $M_{pl,Rd,A}$ and $M_{pl,Rd,B}$ with opposite signs at the end sections A and B of the beam.

Note: $V_{Ed,M} = (M_{pl,Rd,A} + M_{pl,Rd,B})/L$ is the most unfavourable condition, corresponding to a beam with a span L and dissipative zones at both ends.

2.3.2 Columns (referring to Clause 6.6.3 of EC 8)

- The columns shall be verified in compression considering the most unfavourable combination of the axial force and bending moments. N_{Ed}, M_{Ed} and V_{Ed} shall be calculated from the following expressions:

$$\begin{aligned} N_{Ed} &= N_{Ed,G} + 1.1 \times \gamma_{ov} \times \Omega \times N_{Ed,E} \\ M_{Ed} &= M_{Ed,G} + 1.1 \times \gamma_{ov} \times \Omega \times M_{Ed,E} \\ V_{Ed} &= V_{Ed,G} + 1.1 \times \gamma_{ov} \times \Omega \times V_{Ed,E} \end{aligned} \quad (6.6)$$

where

$N_{Ed,G}$, $M_{Ed,G}$, $V_{Ed,G}$ are the compression force, bending moment and shear force in the column due to the non-seismic actions included in the combination of actions for the seismic design situation,
$N_{Ed,E}$, $M_{Ed,E}$, $V_{Ed,E}$ are the compression force, bending moment and shear force in the column due to the design seismic action,
γ_{ov} = the overstrength factor (recommended value = 1.25),
Ω = the minimum value of $\Omega_i = M_{pl,\,Rd,i}/M_{Ed,\,i}$ of all beams in which dissipative zones are located; $M_{Ed,\,i}$ is the design value of the bending moment in beam i in the seismic design situation and $M_{pl,Rd,i}$ is the corresponding plastic resistance moment.

- In columns where plastic hinges form, the resistance moment in those hinges shall be equal to $M_{pl,Rd}$.
- The above plastic moment of resistance of the columns should be verified in accordance with EN 1993-1-1: 2005, EC 3 (Eurocode, 2005).

- The column shear force V_{Ed} resulting from the structural analysis should satisfy the following expression:

$$\frac{V_{Ed}}{V_{pl,Rd}} \leq 0.5. \qquad (6.7)$$

- The transfer of the shear forces from the beam to the columns shall be in accordance with Section 6 of EN 1993-1-1: 2004, EC 8 (Eurocode, 2004).
- The shear resistance of the framed web panels of the beam/column connections (refer to Fig. 6.10 of EC 8) should satisfy the following expression:

$$\frac{V_{wp,Ed}}{V_{wp,Rd}} \leq 1 \qquad (6.8)$$

where

$V_{wp,Ed}$ = design shear force in the web panel,
$V_{wp,Rd}$ = plastic shear resistance of the web panel.

2.3.3 Beam to column connections

- For the structure designed to dissipate energy in the beams, the connections of the beams to the columns should be designed for the required degree of overstrength factor (γ_{ov} = 1.25), taking into consideration the moment of resistance $M_{pl,Rd}$ and the shear force ($V_{Ed,G} + V_{Ed,M}$) already evaluated in Clause 19.3.1.

References

Eurocode, 2004. BS EN 1998-1: 2004, Eurocode 8. Design of Structures for Earthquake Resistance.
Eurocode, 2005. BS EN 1993-1-1: 2005, Eurocode 3. Design of Steel Structures.

CHAPTER 3

Load Combinations of Seismic Action with Other Actions

3.1 Initial effects of the design of seismic action with other actions (referring to Clause 3.2.4 of EC 8)

The initial effects of the design of the seismic action shall be evaluated by taking into account the presence of the masses associated with all gravity loads appearing in the following combination of actions:

$$mg = [\Sigma G_{k,j} + \Sigma \psi_{Ei} \times Q_{k,i}]$$

where

mg = gravity loads of all masses = mass × acceleration in kN, (3.17)

ψ_{Ei} = combination coefficient for variable actions (see Clause 4.2.4), where $\psi_{Ei} = \varphi \times \psi_{2i}$; the recommended values of φ are listed in Table 3.1 and $Q_{k,i}$ = variable load, including snow load,

$G_{k,j}$ = permanent dead load.

3.2 The combination coefficients ψ_{Ei}

The combination coefficients $\psi_{E,i}$ take into account the likelihood of the loads $Q_{k,i}$ not being present over the entire structure during an earthquake. These coefficients may also account for a reduced participation of masses in the motion of the structure due to the non-rigid connection between them.

The combination coefficients ψ_{Ei} introduced in the above equation for the calculation of the effects of the seismic actions shall be computed from the following expression:

$$\psi_{Ei} = \varphi \times \psi_{2i}. \qquad (4.2)$$

The recommended values of φ are given in Table 3.1.

The values of ψ_{2i} for the imposed loads in buildings vary in categories A to E (see EN1991-1-1), as shown in Table 3.2.

In our case, with a category E storage area, referring to Table 3.1 we have $\varphi = 1.0$. Also, referring to Table 3.2, $\varphi_2 = 0.8$. So

$$\psi_{Ei} = \varphi \times \psi_{2i} = 1.0 \times 0.8 = 0.8$$

Table 3.1. Recommended values of φ for calculating ψ_{Ei} (based on Table 4.2 of EC 8, Part 1, BS EN 1998-1: 2004)

Type of variable action	Storey	φ
Categories A–C[a]	• Roof • Stories with correlated occupancies • Independently occupied storeys	1.0 0.8 0.5
Categories D–F[a] and Archives		1.0

[a] Categories as defined in BS EN 1991-1-1: 2002

Table 3.2. Recommended values of ψ factors for buildings (based on Table A1.1 of EN 1990: 2002(E) (Eurocode, 2002))

Actions	ψ_0	ψ_1	ψ_2
Imposed loads in buildings, category (see EN 1991-1-1)			
Category A: domestic residential areas	0.7	0.5	0.3
Category B: office areas	0.7	0.5	0.3
Category C: congregation areas	0.7	0.7	0.6
Category D: shopping areas	0.7	0.7	0.6
Category E: storage areas	1.0	0.9	0.8
Category F: traffic area, vehicle weight ≤ 30 kN	0.7	0.7	0.6
Category G: traffic area, 30 kN < vehicle weight ≤ 160 kN	0.7	0.5	0.3
Category H: roofs	0.0	0.0	0.0
Snow loads on buildings (see EN 1991-1-3)			
Finland, Iceland, Norway, Sweden	0.7	0.5	0.2
Remainder of CEN Member states, for sites located at altitude H > 1000 m a.s.l.	0.7	0.5	0.2
Remainder of CEN Members states, for sites located at altitude H ≤ 1000 m a.s.l.	0.5	0.2	0.0
Wind loads on buildings (see EN 1991-1-4)	0.6	0.2	0.0
Temperature (non-fire) in buildings (see EN 1991-1-5)	0.6	0.5	0.0

The gravity loads are therefore

$$mg = [\Sigma G_{k,j} + 0.8 \times \Sigma Q_{k,i}] \text{ kN}.$$

References

Eurocode, 2002. BS EN 1990: 2002(E), Basis of Structural Design.
Eurocode, 2004. BS EN 1998-1: 2004, Eurocode 8. Design of Structures for Earthquake Resistance.

CHAPTER 4

General Introduction to Vibration and Seismic Analysis

4.1 Design data

We will consider the analysis and design of an industrial steel-framed building (see Figs. 5.1, 5.2 and 5.3 in Chapter 5) in a region subjected to earthquake forces, using the following design data:

- Span (c/c of column) = 16 m.
- Height at eaves level = 10.5 m.
- Height from eaves level to apex = 1.0 m.
- Length of building = 105 m.
- Spacing of main frames = 10.5 m.
- All joints are welded connections.
- The sides and roof are covered by corrugated galvanised steel sheets and supported by roof purlins and side rails that are connected to the main frames.

4.2 Seismological actions (referring to Clause 3.2 of EC 8)

The construction area is in an earthquake zone of *strong* intensity, which is equivalent to a modified Mercalli (MM) scale value of VII–VIII and between 6.2 and 6.8 on the Richter scale. We will assume a horizontal ground acceleration of 0.1g. For most of the application of BS EN 1998-1: 2004 (Eurocode, 2004), the hazard is described in terms of a single parameter a_{gR}, i.e. the value of the reference peak ground acceleration on type A ground (see Table 4.1).

The reference peak ground acceleration, chosen by the National Authorities for each seismic zone, corresponds to the reference return period T_{NCR} of the seismic action for the no-collapse requirement (or equivalently the reference probability of exceedence in 50 years, P_{NCR}) chosen by the National Authorities (see Clause 2.1(1)P, "No-collapse requirement"). An importance factor $\gamma_I = 1.0$ is assigned to this reference period.

Low sensitivity is recommended:

- when the design ground acceleration on type A ground $a_g \leq 0.08$g or those where the product $a_g \times S \leq 0.1g$;

where S is the soil factor. In cases of very low sensitivity, the provisions of BS EN 1998 need not be satisfied.

Table 4.1. Ground types (based on Table 3.1 of EC 8 (Eurocode, 2004))

Ground types	Description of stratigraphic profile	Parameters		
		Shear wave velocity $V_{S,30}$ (m/s)	*NSPT* (blows/ 30 cm)	C_u (kPa)
A	Rock or other rock-like geological formation, including at most 5 m of weaker material at the surface	> 800	-	-
B	Deposits of very dense sand, gravel, or very stiff clay, at least several tens of metres in thickness, characterised by a gradual increase of mechanical properties with depth	360–800	> 50	> 250
C	Deep deposits of dense or medium-dense sand, gravel or stiff clay with thickness from several tens to many hundreds of metres	180–360	15–50	70–250
D	Deposits of loose to medium cohesionless soil (with or without some soft cohesive layers), or of predominantly soft to firm cohesive soil	< 180	< 15	< 70
E	A soil profile consisting of a surface alluvium layer with V_s values of type C or D and thickness varying between about 5 m and 20 m, underlain by stiffer material with V_s > 800 m/s	-	-	-
S1	Deposits consisting, or containing a layer at least 10 m thick, of soft clay/silts with plasticity index (P_I > 40) and high water content	< 100 (indicative)	-	10–20
S2	Deposits of liquefiable soils or sensitive clays, or any other soil profile not included in types A–E or S1	-	-	-

4.3 Identification of ground types

The stratigraphic profiles and geotechnical soil parameters influence the seismic actions. Table 4.1 describes the ground types.

The site should be classified according to the value of the average shear wave velocity $V_{s,30}$, if available. Otherwise the value of *NSTP* should be used in selecting the ground types. In our case, we assume ground type D.

NSPT = Nos of blows/300mm penetration in SPT (Soil Penetration Test) in geotechnical term.

4.4 Basic representation of the seismic action (referring to Clause 3.2.2 of EC 8)

- The earthquake motion at a given point on the surface is represented by an elastic horizontal ground acceleration response spectrum, which is called the *elastic response spectrum*, symbolised by $S_e(T)$. For T (vibration period) equal to zero, the spectral acceleration given by this spectrum equals the design ground acceleration on type A ground, multiplied by the soil factor S.

- The shape of the elastic response spectrum is taken as being the same for two levels of seismic action introduced in Clauses 2.1(1)P and 2.2.1(P)P for the no-collapse requirement (ultimate-limit-state design seismic action) and for the damage-limitation requirement.
- The horizontal seismic action is defined by two orthogonal components assumed to be independent and represented by the same response spectrum.
- For important structures where the importance factor (γ_I) exceeds 1.0, topographic amplification effects should be considered.
- Time–history representations: the seismic motion may also be represented in terms of ground acceleration time histories and related velocity and displacement.

4.5 Horizontal elastic response spectrum

Referring to Clause 3.2.2.2 of EC 8 ("Horizontal elastic response spectrum"), for the horizontal components of the seismic action, the elastic response spectrum $S_e(T)$ is defined by the following expressions:

$$0 \leq T \leq T_B: \quad S_e(T) = a_g \times S \times [1 + T/T_B \times (\eta \times 2.5 - 1] \quad (3.2)$$
$$T_B \leq T \leq T_C: \quad S_e(T) = a_g \times S \times 2.5 \quad (3.3)$$
$$T_C \leq T \leq T_D: \quad S_e(T) = a_g \times S \times \eta \times 2.5[T_C/T] \quad (3.4)$$
$$T_D \leq T \leq 4s: \quad S_e(T) = a_g \times S \times \eta \times 2.5[T_C \times T_D/T^2] \quad (3.5)$$

where

$S_e(T)$ = elastic response spectrum,
T = vibration period of a linear single-degree-of-freedom system,
a_g = design ground acceleration on type A ground ($a_g = \eta \times a_{gR}$),
T_B = lower limit of the period of the constant spectral acceleration branch,
T_C = upper limit of the period of the constant spectral acceleration branch,
T_D = the value defining the beginning of the constant displacement response range of the spectrum,
S = the soil factor,
η = the damping correction factor with a reference value of $\eta = 1$ for 5% viscous damping.

The values of the periods T_B, T_C and T_D and of the soil factor S describing the shape of the elastic response spectrum depend upon the ground type. If the earthquake that contributes most to the seismic hazard defined for the site, for the purpose of probabilistic hazard assessment two types of spectra are recommended: Type 1 and Type 2.

(1) **Type 1:** If the surface-wave earthquake magnitude $M_s > 5.5$, it is recommended to use a Type 1 spectrum as shown in Table 4.2.
(2) **Type 2:** If the surface-wave earthquake magnitude $M_s < 5.5$, it is recommended to adopt a Type 2 spectrum (see Table 4.3).

In our case $M_s > 5.5$, so we choose the values in Table 4.2 for ground type D, i.e.

$S = 1.35$, $T_B(s) = 0.2$, $T_C(s) = 0.8$ and $T_C(s) = 2.0$.

If the surface-wave magnitude $M_s = 5.5$ the Type 2 spectrum should be adopted (recommended).

Table 4.2. Values of S, T_B, T_C and T_D for a Type 1 spectrum (based on Table 3.2 of EC 8)

Ground type	S	T_B(S)	T_C(S)	T_D(S)
A	1.0	0.15	0.4	2.0
B	1.2	0.15	0.5	2.0
C	1.15	0.20	0.6	2.0
D	1.35	0.20	0.8	2.0
E	1.4	0.15	0.5	2.0

Table 4.3. Values of S, T_B, T_C and T_D for a Type 2 spectrum (based on Table 3.3 of EC 8)

Ground type	S	T_B(s)	T_C(s)	T_D(s)
A	1.0	0.05	0.25	1.2
B	1.35	0.05	0.25	1.2
C	1.5	0.10	0.25	1.2
D	1.8	0.10	0.30	1.2
E	1.6	0.05	0.25	1.2

4.6 Importance classes and importance factors (referring to Clause 4.2.5 of EC 8)

Referring to Clause 4.2.5 of EC 8:

- Buildings are organised into four important classes, depending on the consequences of collapse of human life, on their importance for public safety and civil protection in the immediate post-earthquake period, on the social consequences of collapse and on the economic consequences of collapse.
- The importance classes are characterised by different *importance factors* γ_I. An importance factor is assigned to each importance class. Wherever feasible, this factor should be derived so as to correspond to a higher or lower value of the return period of the seismic event (with regard to the reference return period) as appropriate for the design of the specific category of structure (referring to Clause 3.2.1(3)).

  The probability of return period of 50 years is generally assumed. The value of the importance factor depends on the seismic zone of the country, on seismic hazard conditions and on public safety considerations (Clause 4.2.5(5)P). The recommended values of γ_I for different importance classes are given in Table 4.4.

In our case, the building is classified as an industrial building of vital importance, because this structure lies on the critical path in the production of an industrial final product in a processing plant. Therefore, we assume an importance class = IV, with an importance factor of $\gamma_I = 1.4$. For an importance class IV building and ground type D, the product $a_g S$ is then

$$\begin{aligned} a_g S &= [a_g \times \gamma_I \text{ (importance factor } \gamma_I \text{ of importance class IV)} \times S \text{ (soil factor of ground type D)}] \times g \\ &= [0.10 \times 1.4 \times 1.35]g \\ &= 0.189g > 0.1g. \end{aligned}$$

As this value is greater than 0.1g the building will be designed with *high sensitivity*.

Table 4.4. Importance classes and importance factors for buildings (based on Table 4.3 of EC 8)

Importance class	Buildings	Importance factor (γ_I)
I	Buildings of minor importance for public safety, e.g. agricultural buildings, etc.	0.8
II	Ordinary buildings, not belonging to any other category.	1.0
III	Buildings whose seismic resistance is of importance in view of the consequences associated with the collapse, e.g. schools, assembly halls, cultural institutions, etc.	1.2
IV	Buildings whose integrity during an earthquake is of vital importance for civil protection, e.g. hospitals, fire stations, power plants, etc.	1.4

4.7 Determination of the fundamental period of vibration

The *fundamental period of vibration* T_1 may be calculated by the methods of structural dynamics (for example the Rayleigh method). For buildings with maximum height limited to 40 m, the fundamental period of vibration may be given by the following expression:

$$T_1 = C_1 \times H^{0.75} \tag{4.6}$$

where

C_1 = a factor for the moment of resistance
= 0.085 for moment resistance space steel frames
= 0.075 for concrete frames and eccentrically braced steel
= 0.05 for all other structures,
H = height of the building in m from the foundation or from the top of a rigid basement.

In our case:

- The building under consideration is of moment-resisting space steel-framed construction, so $C_1 = 0.085$.
- The height of the building H up to the apex is 11.5 m < 40 m, so the above expression (4.6) is applicable.

Thus

$T_1 = 0.085 \times 11.5^{0.75} = 0.53$ sec < 2.0 sec. Satisfactory

4.8 Computation of the horizontal elastic response spectrum (referring to Clause 3.2.2.2 of EC 8)

For the horizontal components of the seismic action, the elastic response spectrum $S_e(T)$ is computed from the following expression:

$$T_B \leq T \leq T_C: \quad S_e(T) = a_g \times S \times \eta \times 2.5 \tag{3.3}$$

Table 4.5. Recommended values of parameters describing the vertical elastic response spectrum (based on Table 3.4 of EC 8)

Spectrum	a_{vg}/a_g	T_B (S)	T_C(S)	T_D(S)
Type 1	0.90	0.05	0.15	1.0
Type 2	0.45	0.05	0.15	1.0

where

$T = 0.53$ sec is greater than $T_B = 0.2$ sec,
$T = 0.53$ sec is less than $T_C = 0.8$ sec

(see Table 4.2). So, the expression (3.3) is applicable. The horizontal elastic response spectrum is therefore calculated to be

$$S_e(T) = 0.1 \times 1.35 \times 1.0 \times 2.5 = 0.34g$$

where $\eta = 1.0$ (assumed).

4.9 Computation of the vertical elastic response spectrum (referring to Clause 3.2.2.3 of EC 8)

The vertical component of the seismic action shall be represented by an elastic response $S_{ve}(T)$ and is computed from:

$$T_B \leq T \leq T_C: \qquad S_{ve}(T) = a_{vg} \times \eta \times 3.0 \qquad (3.9)$$

where

$T = 0.53$ sec is greater than $T_B = 0.2$ sec,
$T = 0.53$ sec is less than $T_C = 0.8$ sec

(see Table 4.2). So, the above expression (3.9) is applicable in this case.
The recommended values of the parameters describing the vertical spectrum for a Type 1 or Type 2 spectrum are given in Table 4.5, where

a_{vg} = design ground acceleration in the vertical direction,
a_g = design ground acceleration on type A ground.

In our case, since M_s is greater than 5.5, we shall adopt *type 1*. Thus,

$$a_{vgr} = [a_{vg}/a_g \times a_g \times \text{importance factor } \gamma_1 \text{ for class IV}] \times g$$
$$= (0.9 \times 0.1 \times 1.4)g = 0.126g.$$

Referring to Clause 4.3.3.5.2 (of EC 8), if a_{vgr} is greater than 0.25g, the vertical component of the seismic action as defined in Clause 3.2.2.3 (of EC 8) should be taken into account. In our case, a_{vgr}= (0.126g) is less than 0.25g. Therefore, the vertical component of seismic action shall be neglected.

References

Eurocode, 2004. BS EN 1998-1: 2008, Eurocode 8. Design of Structures for Earthquake Resistance.

CHAPTER 5

HBI (Heavy Briquette Iron) Storage Building

5.1 Brief description of the building

The HBI storage building is of steel-framed construction for temporarily storing heavy briquette iron (HBI), and for transferring the material to the melting shop building through underground hoppers and a conveyor system.

The building is a single-bay rigid-framed structure spanning 16 m between the legs of the frame, and 105 m long covering the whole length of the stock pile of HBI material. The spacing of the main frame is 10.5 m. The bases of the frames are considered hinged, and subjected to all vertical and horizontal loadings, including seismic forces.

A small extension at mid length of the building is provided to serve as an annexe. The annexe frames are assumed to be connected by hinges to the main buildings, and the bases are assumed hinged.

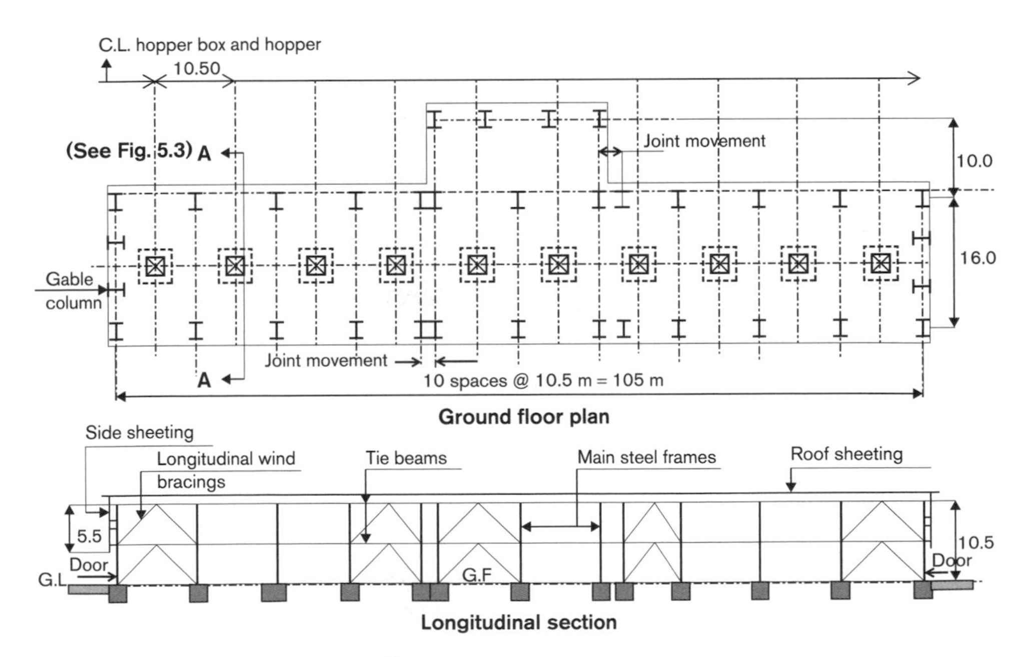

Fig. 5.1. Ground floor plan of the HBI storage building

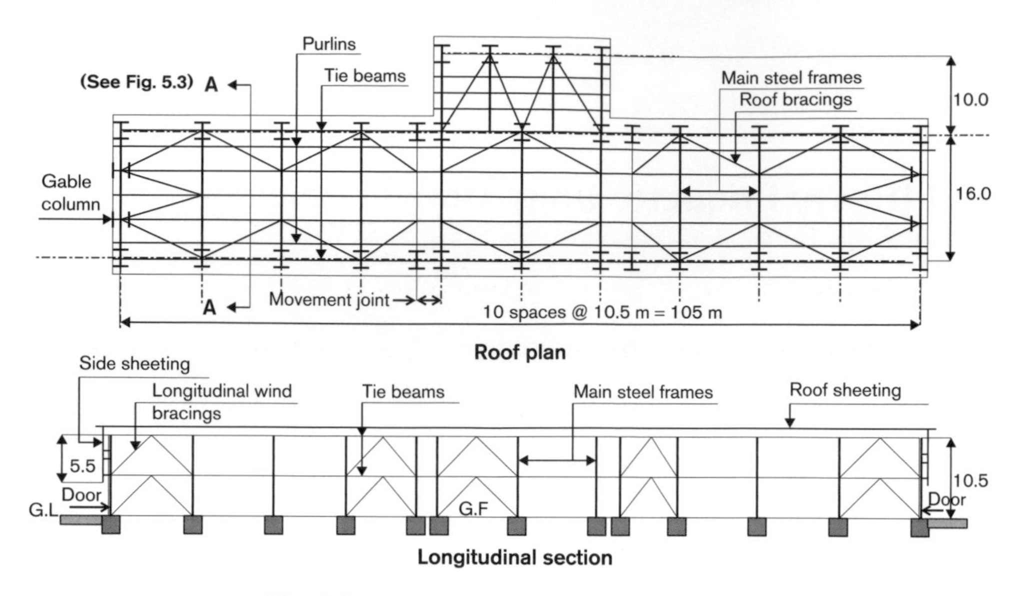

Fig. 5.2. Roof plan of the HBI storage building

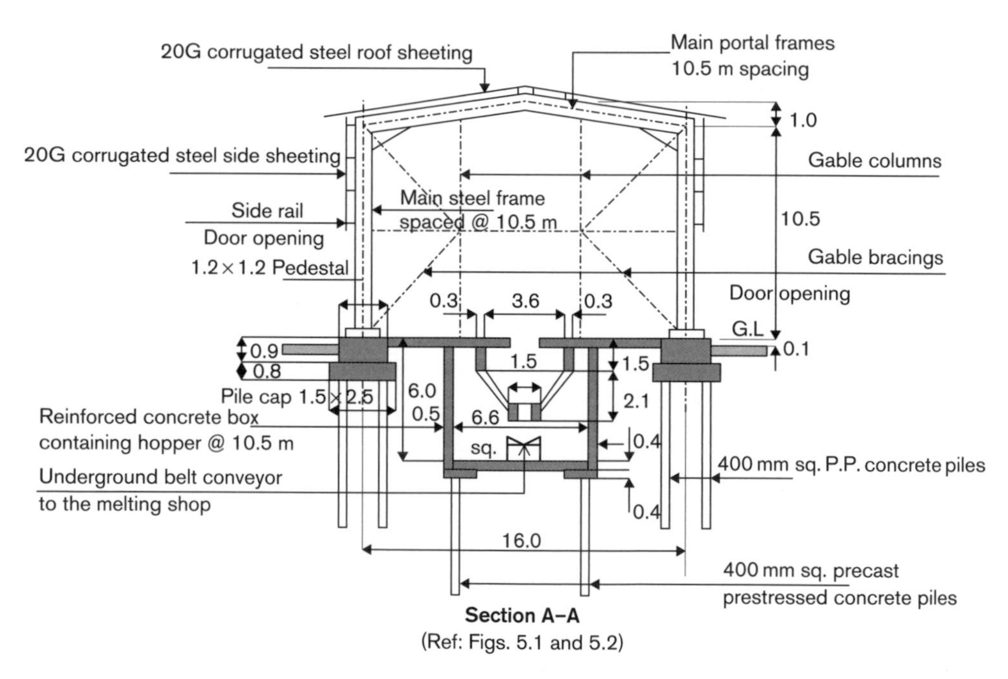

Fig. 5.3. Section A–A of the HBI storage building

An overhead conveyor system, which runs along the length of the building, is suspended from the roof frames inside the structure. This conveyor system transports the HBI material from outside and temporarily deposits the material inside the building. The building also houses a series of underground hopper box structures with hoppers inside the box constructed along the length of the building. These hoppers transfer the stored material to the melting shop building by means of an underground and overground belt conveyor system (see Figs. 5.1, 5.2 and 5.3).

5.2 Design philosophy

In accordance with EC 8 (Eurocode, 2004), two fundamental seismic design levels have been considered:

(1) **No-collapse requirement (referring to Section 2.1(1)P of EC 8):** The structure shall be designed and constructed in such a way that the structure shall withstand the designed seismic action as defined in Section 3 (EC 8) so that the overall structural integrity and the residual load-bearing capacity have been maintained after the occurrence of an earthquake.

To satisfy the above requirements the structure needs to be designed by the ultimate-limit-state (ULS) method. "No-collapse" corresponds to seismic action based on a recommended probability of exceedance P_{NCR} of 10% in 50 years, or a return period T_{NCR} of 475 years.

(2) **Damage-limitation requirement:** The structure shall be designed and constructed to withstand a seismic action having a larger probability of occurrence than the design seismic action, without the occurrence of damage and associated limitations of use, the costs of which would be disproportionately high in comparison with the costs of the structure itself.

"Damage-limitation requirement" relates to a recommended probability P_{NCR} of 10% in 10 years, or a return period T_{NCR} of 95 years. Also, to satisfy the above requirements, the structure shall be analysed and designed by the serviceability-limit-state (SLS) method to limit against unacceptable damage and maintain an adequate degree of reliability.

Reliability differentiation (referring to Section 2.1(3)P) is carried out by classifying structures into different importance classes. An importance factor γ_I is assigned to each importance class. For each seismic zone, with the return period T_{NCR} of the seismic action for the no-collapse requirement or equivalently the reference probability of exceedance in 50 years P_{NCR}, an importance factor $\gamma_I = 1.0$ is generally assigned.

In order to satisfy the fundamental requirements as mentioned above, the following limit states shall be checked:

- Ultimate limit states (ULS).
- Damage-limitation states (in serviceability limit states (SLS)).

5.2.1 Ultimate limit states

Ultimate limit states are those associated with collapse or other forms of structural failure which might endanger the safety of people. In the ultimate-limit-state concept, the structure should be designed to a limiting stage value beyond which the structure becomes unfit for intended use.

In order to limit the uncertainties and to retain adequate structural resistance and energy dissipation capacity under seismic actions, the following specific measures in the design should be taken:

- The structure shall be of a simple and regular form, both in plan and in elevation.
- In order to ensure an overall dissipative and ductile behaviour, a brittle failure or the premature formation of unstable mechanisms shall be avoided.

- The detailing of critical regions shall be such as to maintain the capacity to transmit the necessary forces and dissipate energy under cyclic conditions.
- The stiffness of the foundations shall be adequate for transmitting the actions received from the superstructure to the ground as uniformly as possible.

The building shall be analysed and designed to resist the earthquake forces in the project site based on non-dissipative or dissipative behaviour. The non-dissipative behaviour implies a largely elastic response and is normally limited to areas of low seismicity or to structures of special importance. Generally, in most of the structures, an economical dissipative design is carried out whereby significant plastic deformations can be accommodated under extreme circumstances by assigning a behaviour factor to reduce the code-specified lateral forces resulting from an idealised elastic response spectrum. The above procedure is performed in combination with capacity design approaches, requiring the determination of a pre-defined plastic mechanism, coupled with the provision of adequate ductility in plastic zones and appropriate strength in other regions.

Referring to Clause 4.3.3 of EC 8 (method of analysis), "The lateral force method of analysis" for the welded steel portal framed building "on the basis of the linear-elastic behaviour" meeting the conditions as given in Clause 4.2.3 (criteria for structural regularity) has been adopted.

- **Criteria for regularity in plan**
  The slenderness ratio is defined to be $\lambda = L_{max}/L_{min} \leq 4$, where L_{max} = larger dimension in plan and L_{min} = smaller dimension in plan.
  In our case, the total length of the building = 105 m and the width of the building = 16 m. We have divided the total length (105 m) of the building into three separate independent portions: two external units of 42 m = 84 m; and a central unit of 21 m. A clear gap between the units has been left so that the earthquake force on each unit will not affect the adjacent one.
  Therefore, $\lambda = 42/16 = 2.65 < 4$. So, the building satisfies the criteria for regularity in plan.
- **Criteria for regularity in elevation**
  Referring to Fig 5.3, we can see that in plan the shape of the building is uniform except the central two bays. To satisfy the regularity criteria the whole length is divided in three sections - namely two external portions consisting of 4 bays and the central portion of 2 bays. These three portions are separated by contraction joints so that during earthquakes these segments have got adequate strength and stability to resist the seismic forces and are designed accordingly.
- **Criteria for well-distributed and relatively rigid cladding**
  The criteria is satisfied by providing the building with well-distributed rigid cladding on the roof and vertical sides.

References

Eurocode, 2004. BS EN 1998-1: 2004, Eurocode 8. Design of Structures for Earthquake Resistance.

CHAPTER 6

Computation of the Design Base Shear Force in an Elastic Analysis

6.1 Design spectrum of an elastic analysis

Referring to Clause 3.2.2.5 (of EC 8), in order to avoid explicit inelastic structural analysis in the design, the capacity of the structure to dissipate energy, through mainly ductile behaviour of its elements and/or other mechanisms, is taken into account by performing an elastic analysis based on a response spectrum reduced with respect to the elastic one, henceforth called a *design spectrum* $S_d(T)$. This reduction is accomplished by introducing the behaviour factor "q".

6.1.1 Behaviour factor

Recall, the *behaviour factor q* is an approximation of the ratio of the seismic forces that the structure would experience if its response was completely elastic with 5% viscous damping, to the seismic forces that may be used in the design with a conventional elastic model, still ensuring a satisfactory response of the structure. The values of the behaviour factor q, which also takes into account the influence of the viscous damping being different from 5%, are given for various materials and structural systems according to the relevant ductility classes in the various Parts of BS EN 1998 (Eurocode, 2004).

Referring to Section 6 of EC 8, steel-framed buildings may be designed based on non-dissipative or dissipative behaviour. The non-dissipative behaviour implies largely elastic response and is normally limited to areas of low seismicity or to structures of special importance. It is economical to design based on dissipative behaviour so that significant plastic deformations can be accommodated under extreme conditions.

In most cases, steel structures are designed based on dissipative behaviour by assigning a behaviour factor to reduce the code-specified lateral forces resulting from an idealised elastic response spectrum. This is carried out in conjunction with capacity design approaches, requiring the determination of a pre-defined plastic mechanism coupled with the provision of adequate ductility in plastic zones and appropriate strength in other regions.

The adoption of the behaviour factor q enables the use of standard elastic analysis tools for the seismic design of regular structures, using a set of reduced forces. Reference values of the behaviour factor are given in Table 6.1, which is based on Table 6.1 of EC 8.

Referring to Clause 6.1.2(1P) (of EC 8), earthquake-resistant steel-framed buildings shall be designed in accordance with one of the following concepts (see Table 6.1).

- **Concept a:** Low-dissipative structural behaviour
- **Concept b:** Dissipative structural behaviour

Table 6.1. Design concepts, structural ductility classes and upper limit reference values of the behaviour factor (based on Table 6.1 of EC 8)

Design concept	Structural ductility class	Range of reference values of the behaviour factor q
Concept a: Low-dissipative structural behaviour	DCL (ductility class low)	$\leq$ 1.5–2
Concept b: Dissipative structural behaviour	DCM (ductility class medium)	$\leq$ 4 Also limited by the values of Table 6.2
	DCH (ductility class high)	Only limited by the values of Table 6.2

Table 6.2. Upper limit of reference behaviour factors for regular moment-resisting and concentrically braced frames (based on Tables 6.2 and 6.3 of EC 8)

Structural type	Ductility class	Behaviour factor q	Remarks
Non-dissipative regular structure of low seismicity	DCL	1.5	Detailed to EC 3 requirements only
Dissipative regular moment-resisting frames	DCH	$5 \times \alpha_u/\alpha_1$	Recommended α_u/α_1 single portal frame = 1.1 single span multistorey = 1.2 multi-span multistorey = 1.3 $\alpha_u/\alpha_1 \leq 1.6$
Note: dissipative zones mainly in beams	DCH	4	Cross-section in dissipative zones: DCM ($1.5 < q \leq 2$) = class 1, 2 or 3
	DCM	4	DCM ($2.0 < q \leq 4$) = class 1 or 2 DCH ($q > 4$) = class 1
Dissipative regular braced frames	DCH	4	Cross-section in dissipative zones same as above
Note: dissipative zone mainly in tension diagonals	DCM	4	

For regular structures in areas of low seismicity, a value of q equal to 1.5 may be adopted without applying dissipative procedures, recognising the presence of inherent overstrength and ductility. In this case, the structure is classified as DCL, for which global elastic analysis can be utilised, and the resistance of members and connections designed in accordance with EC 3 (Eurocode, 2005).

When the value of q is greater than 1.5, the structure shall have sufficient ductility and resistance within the dissipative zones. Table 6.2 gives the range of behaviour factors for regular moment-resisting and concentrically braced frames.

In our case, we assume a dissipative moment-resisting single-portal frame. Thus, from the Table 6.2, we select the behaviour factor with ductility class DCM and with cross-section class 1 or 2 and $q = 2.5$.

6.1.2 Design spectrum $S_d(T)$

Referring to the following expressions in EC 8 for $S_d(T)$:

$$0 \le T \le T_B: \quad S_d(T) = a_g \times S \times \left[\frac{2}{3} + \frac{T}{T_B} \times \left(\frac{2.5}{q} - \frac{2}{3} \right) \right] \tag{3.13}$$

$$T_B \le T \le T_C: \quad S_d(T) = a_g \times S \times \left[\frac{2.5}{q} \right]. \tag{3.14}$$

$$T_C \le T \le T_D: \quad S_d(T) = a_g \times S \times \left[\frac{2.5}{q} \right] \times \left[\frac{T_C}{\mathrm{T}} \right] \tag{3.15}$$

$$\text{but should be} \ge \beta \times a_g$$

$$T_D \le T: \quad S_d(T) = a_g \times S \times \left[\frac{2.5}{q} \right] \times \left[T_C \times \frac{T_D}{T^2} \right] \tag{3.16}$$

$$\text{but should be} \ge \beta \times a_g$$

where β = lower bound factor for the horizontal design spectrum (recommended value = 0.2), and referring to Table 4.2 (Table 3.2 of EC 8), with ground type D,

$S = 1.35$; $T_B = 0.2$; $T_C = 0.8$; $T_D = 2.0$

and already calculated in Chapter 4,

$T_1 = T = 0.53$ sec $> T_B = (0.2)$ but less than $T_C = (0.8)$,

the expression that applies in our case is

$$T_B \le T \le T_C: \quad S_d(T) = a_g \times S \times \left[\frac{2.5}{q} \right]. \tag{3.14}$$

Thus, the design spectrum is calculated to be

$$S_d(T) = 0.1 \times 1.35 \times \frac{2.5}{2.5} = 0.14g$$

6.2 Computation of the design seismic base shear force

Referring to Section 4 of EC 8 (Eurocode, 2005), "Design of buildings in seismic regions", the aspect of a seismic hazard shall be taken into consideration in the early stages of conceptual design, thus enabling the achievement of a structural system which, within acceptable costs, satisfies the fundamental requirements specified in Clause 2.1 (EC 8). The main guiding principles governing this conceptual design are:

- **Structural simplicity:** The structural members should be arranged in a simple system without using redundancies, if possible, so that the seismic forces are transmitted in clear and direct paths, and the modelling, analysis, detailing and construction of simple structures become easy, and at the same time the prediction of seismic behaviour in structural members could be reliably ascertained.
- **Uniformity and symmetry:** The uniformity in plan by an even distribution of the structural elements will ensure short and direct transmission of the inertia forces created in the distributed masses of the building. If necessary, uniformity may be idealised by subdividing the entire building's seismic joints into dynamically independent units.

In our case, we have introduced movement joints to act as dynamically independent units as shown in Figs. 5.1 and 5.2 (see Chapter 5).

6.3 Design criteria for structural regularity

Referring to Clause 4.2.3.1 (EC 8):

- For the purpose of seismic design, building structures are categorised into being *regular* or *non-regular*.
- With regard to the implications of structural regularity on analysis and design, separate consideration is given to the regularity characteristics of the building in plan and in elevation as shown in Table 6.3 (Table 4.1 of EC 8).

Referring to Clause 4.3.3.2.1(2), the requirement of this is deemed to be satisfied in buildings which fulfil both of the following conditions:

- They have fundamental periods of vibration T_1 in the two main directions in buildings which are smaller than the values:

$$4 \times T_C \text{ and } 2.0 \times S \qquad (4.4)$$

- They meet the criteria for regularity in elevation given in Subclause 4.2.3.3.

Table 6.3. Consequences of structural regularity on seismic analysis and design (based on Table 4.1 of EC 8)

	Regularity		Allowed simplification		
Case	Plan	Elevation	Model	Linear-elastic analysis	Behaviour factor (for linear analysis)
1	Yes	Yes	Planar	Lateral force	Reference value
2	Yes	No	Planar	Model	Decreased value
3	No	Yes	Spatial	Lateral force	Reference value[a]
4	No	No	Spatial	Model	Decreased value

[a] Under the specific conditions given in Subclause 4.3.3.1(8);

- the building shall have well-distributed and relatively rigid cladding and partitions;
- the building height shall not exceed 10 m.

For our building, Case 1 in Table 6.3 is satisfied, so a lateral force method of linear-elastic analysis shall be adopted. Thus, we have:

- **Referring to Clause 4.2.3.1(1)P:** The building is designed as regular.
- **Referring to Clause 4.2.3.1(3)P:** Case 1 in Table 6.3 is satisfied.
- **Referring to Clause 4.2.3.1(4):** Criteria describing regularity in plan and in elevation are given in Clauses 4.2.3.2 and 4.2.3.3.
- **Referring to Clause 4.2.3.1(5)P:** The regularity criteria given in Clauses 4.2.3.2 and 4.2.3.3 should be taken as necessary conditions. It shall be verified that the assumed regularity of the building structure is not impaired by other characteristics.
- **Referring to Clause 4.2.3.1(6):** Reference values of the behaviour factor are given in Section 6 ("Specific rules for steel-framed buildings"). See Table 6.1 and Table 6.2.
- The behaviour factor q is assumed to be equal to 2.5 for dissipative structural steel-framed buildings for ductility class (DCM).

6.3.1. Criteria for regularity in plan (referring to Clause 4.2.3.2)

- **Referring to Clauses 4.2.3.2(1)P to 4.2.3.2(4):**
  The building satisfies the criteria for regularity in plan (see Figs. 5.1, 5.2 and 5.3 in Chapter 5).
- **Referring to Clause 4.2.3.2(5):**
  As seen earlier in Chapter 5, the slenderness ratio $\lambda = L_{max}/L_{min}$ of the building in plan shall not be greater than 4 where

  L_{max} = larger dimension in plan = 42 m length of building,
  L_{min} = smaller dimension in plan = 16 m = span of frame.

  In order to satisfy the above condition:

  total length of building = 105 m,
  width of building = 16 m.

  We have divided the total length of the building into three separate independent portions: two external portions, each unit of 42 m × 2 = 84 m; and the central unit of 21 m. A clear gap has been left between units so that the seismic force on an individual unit will not affect the adjacent one.

  So, in our case, $\lambda = 42/16 = 2.62 < 4$, which satisfies the criteria for slenderness in plan.

- **Referring to Clause 4.2.3.2(6):**
  At each level and for each direction of analysis x and y, the structural eccentricity e_o and the torsional radius r shall be in accordance with the two conditions below, which are expressed for the direction of analysis y:

$$e_{ox} \leq 0.30 \times r_x \quad (4.1a)$$

$$r_x \geq l_s \quad (4.1b)$$

  where

  l_s = radius of gyration of the floor mass in plan.

  The structural eccentricity $e_{ox} = 0$, so it satisfies the criteria for regularity in plan.

- **Referring to Clauses 4.2.3.2(3) and 4.2.3.2 (4):**
  The plan configuration is compact (as shown in Figs. 5.1 and 5.2 in Chapter 5). The in-plan stiffness is achieved by providing well-distributed and relatively rigid cladding on the roof and sides of the building.

6.3.2 Criteria for regularity in elevation (referring to Clause 4.2.3.3)

Referring to Figs. 5.1, 5.2 and 5.3 in Chapter 5, the building is a regular single-storey and single-bay portal-framed steel construction. So, all criteria for regularity in elevation are satisfied.

6.4 Structural analysis

- **Structural model**
  The structural model should be prepared to represent the distribution of stiffness and mass in the building, so that all significant deformation shapes and inertia forces are properly taken into consideration under the seismic action. The structural model must consist of horizontal and vertical members to resist the vertical and horizontal loads.
- **Accidental torsional effects (referring to Clause 4.3.2)**
  Referring to Clause 4.3.2(1)P, in order to account for uncertainties in the location of masses and in the spatial variation of the seismic motion, the calculated centre of mass of each floor *i* shall be considered as being displaced from its nominal location in each direction by an accidental eccentricity:

$$e_{ai} = \pm 0.05 \times L_1 \tag{4.3}$$

where

e_{ai} = accidental eccentricity of storey mass *i* from its nominal location, applied in the same direction on all floors,
L_1 = floor dimension perpendicular to the direction of the seismic action
= 42 m (see floor plan in Fig. 5.1, Chapter 5). The total length of the building is divided into three individual independent parts to avoid the transfer of seismic forces to adjacent parts.

So, the design seismic load is to be increased due to accidental eccentricity by a factor of

$$e_{ai} = \pm 0.05 \times 42 = 2.1.$$

6.5 Method of analysis (referring to Clause 4.3.3 of EC 8)

The seismic effects, as well as the effects of the other actions included in the seismic design situation, may be determined on the basis of the linear-elastic behaviour of the structure.

6.5.1 Linear-elastic method of analysis

Based on the linear-elastic behaviour of the structure, the analysis shall be carried out by a linear-elastic method. Seismic effects and the effects of the other actions included in the seismic design situation will be determined using a linear-elastic model of the structure and the design spectrum given in Clause 3.2.2.5 of EC 8. There are two types of linear-elastic method of analysis that may be used:

(1) Lateral force method of analysis.
(2) Modal response spectrum analysis.

In our case, we shall adopt the lateral force method of analysis.

Modal response spectrum is applied when lateral force methods of analysis can not be applied when response of all modes of vibration contributing significantly to the global response are taken into account. This method is highly complicated where sum of the effective modal masses for the modes taken into account amounts to at least 90% of the total mass of the structure.

6.5.2 Lateral force method of analysis

The *lateral force method of analysis* may be applied to buildings provided that the response to the building is not significantly affected by contributions from modes of vibration higher than the fundamental mode in each direction (see Clause 4.3.3.1 of EC 8). This requirement shall be considered satisfied if the following two conditions are fulfilled:

(1) The building shall have fundamental periods of vibrations T_1 in the two main directions, which will be smaller than the values given in the following expression:

$$T_1 \leq [4 \times T_C] \tag{4.4}$$

and

$$T_1 \leq [2\,\text{sec}].$$

In our case, $4 \times T_C = 4 \times 0.8 = 3.2$ (the value of T_C is obtained from Table 6.2 (or Table 6.2 and Table 6.3 of EC 8) and $T_1 = 0.53$ sec was previously calculated (see Chapter 4). Thus, $T_1 < 4 \times T_C < 2$ sec, and so it satisfies the condition.

(2) Criteria for regularity in elevation. In Chapter 5 we established that the building has regularity in elevation with well-distributed and relatively rigid cladding. Thus, it satisfies the criteria and hence the lateral force method of analysis is applicable.

For buildings satisfying all the above conditions, a linear-elastic analysis using two planar models, one for each main horizontal direction, may be performed. In such cases, however, all seismic action effects resulting from the analysis should be multiplied by a seismic action effect factor $m = 1.25$ (see "Method of analysis", Clause 4.3.3.1(9) of EC 8). So, the design seismic load is to be increased by a factor $m = 1.25$.

6.6 Seismic base shear force

Referring to Clause 4.3.3.2.2 (EC 8), the seismic base shear force F_b for each horizontal direction in which the building is analysed shall be determined using the following expression:

$$F_b = S_d(T_1) \times m \times \lambda \tag{4.5}$$

where

$S_d(T_1)$ = ordinate of the design spectrum (see Clause 3.2.2.5) at T_1,

T_1 = fundamental period of vibration of the building for lateral motion in the direction considered,
m = total mass of the building above the foundation,
λ = correction factor
= 0.85 if $T_1 \leq 2T_C$ and the building has more than two storeys, or
= 1 otherwise.

In our case, $S_d(T_1) = 0.14g$ (already calculated earlier in this chapter) and $T_1 = 0.53$ which is less than $2 \times 0.8 = 1.6$. The building has a single storey, so $\lambda = 1.0$. The base shear force is therefore

$$F_b = 0.14g \times m \times 1.0 = 0.14\ mg.$$

6.7 Design base shear force

The design base shear force is calculated from the equation

$$F_{bd} = F_b \times \text{factor } e_{ai} \times \text{factor } m$$
$$= 0.14\ mg \times \text{factor } e_{ai} \times \text{factor } m$$

where

factor e_{ai} = accidental eccentricity = 2.1 (previously calculated),
factor m = seismic action effect = 1.25 (see Clause 4.3.3).

Therefore, we have

$$F_{bd} = 0.14\ mg \times 1.25 \times 2.1 = 0.37\ mg.$$

The length of the building considered is 42 m, so the design base shear force per unit length F_{bd}/m = 0.37 mg/42 = 0.009 mg/m length.

References

Eurocode, 2004. BS EN 1998-1: 2004, Eurocode 8. Design of Structures for Earthquake Resistance.
Eurocode, 2005. BS EN 1993-1-1: 2005, Eurocode 3. Design of Steel Structures.

CHAPTER 7

Seismic Analysis of Beams and Columns, and Design to Eurocode 3

7.1 Design considerations

Referring to Fig. 5.2 in Chapter 5 showing the longitudinal bracing system along the column lines, the whole length of building is divided into three sections in order to limit the length of each section to 40 m to satisfy the condition as stipulated in the earthquake code of practice EC 8 (Eurocode, 2004). A separation (movement) joint between each section has been introduced so that each section acts independently.

Each section is provided with two bracing systems at each end bay. The function of the bracing system is to resist the wind forces on the gable end and also the earthquake forces assumed to be acting simultaneously.

7.2 Loadings

7.2.1 Characteristic wind loads (W_k)

The wind load (WL) calculations are carried out based on EC 1: Part 1-4 (Eurocode, 2006). All expressions, clauses and references that are quoted hereinafter refer to this code unless stated otherwise.

7.2.1.1 Basic wind velocity

We assume a basic wind velocity of vb = 24 m/s.

7.2.1.2 Mean wind velocity

The mean wind velocity v_m at a height z above ground depends on the terrain roughness and orography and may be calculated from the expression:

$$v_m(z) = v_b \times c_r(z) \times c_o(z) \tag{4.3}$$

where

$c_o(z)$ = orography factor (recommended value = 1.0),

$c_r(z)$ = roughness factor, which accounts for the variability of the mean wind velocity at the site of the structure due to the height above ground and the ground roughness of the terrain upwind of the structure in the wind direction considered.

The recommended procedure for determining $c_r(z)$ at a height z is given by the following expression:

$$c_r(z) = \text{roughness factor} = k_r \times \ln(z/z_o) \quad \text{for } z_{min} \leq z \leq z_{max} \tag{4.4}$$
$$= c_r(z_{min}) \quad \text{for } z \leq z_{min}$$

where

z = height of building above ground level = 11.5 m (see Fig. 5.3 in Chapter 5),
z_o = roughness length,
z_{max} = maximum height,
k_r = terrain factor depending on the roughness length z_o, calculated using the following expression:

$$k_r = 0.19(z_o/z_{o,II})^{0.07} \tag{4.5}$$

where

$z_{o,II}$ = 0.05 m (terrain category II (defined in Table 7.1)),
z_{min} = minimum height (defined in Table 7.1).

If we assume a terrain category II, with

z_{min} = 2 m and z_{max} = 200 m

i.e.

z_o = 0.05,
$k_r = 0.19 \times [z_o/z_o, \text{II}]^{0.07}$, (4.5)
z_{min} = minimum height for terrain category II (see Table 7.1),
z_{max} = to be take n equal to 200 m.

then we have

$$k_r = 0.19 \times [0.05/0.05]^{0.07} = 0.19.$$

Therefore

$$c_r(z) = 0.19 \times \ln(z/z_o) = 0.19 \times \text{In}(11.5/0.05) = 1.03.$$

Table 7.1. Terrain categories and terrain parameters (based on Table 4.1 of EC 1: Part 1-4)

	Terrain category	z_o (m)	z_{min} (m)
0	Sea or coastal area exposed to the open sea	0.003	1
I	Lakes or flat and horizontal area with negligible vegetation and without obstacles	0.01	1
II	Area with low vegetation such as grass and isolated obstacles (e.g. trees, buildings) with separations of at least 20× obstacle height	0.05	2
III	Area with regular cover of vegetation or buildings with isolated obstacles with separations of maximum 20× obstacle height (such as villages, suburban terrain, permanent forest)	0.3	5
IV	Area in which at least 15% of the surface is covered with buildings and their average height exceeds 15 m	1.0	10

The orography factor $c_o(z)$ (e.g. hills, cliffs, etc.) increases the wind velocity but the effects may be neglected when the average slope of the upwind terrain is less than 3°. The upwind terrain may be considered up to a distance of ten times the height of the isolated orographic feature.

In our case, assume the average slope of the upwind terrain is less than 3°, so $c_o(z) = 1.0$. The mean wind velocity is therefore

$$v_m(z) = v_b \times c_r(z) \times c_o(z) = 24 \times 1.03 \times 1.0 = 24.7 \text{ m/s}.$$

7.2.1.3 Peak velocity pressure

The peak velocity pressure $q_p(z)$ at a height z is given by the following expression:

$$q_p(z) = c_e(z) \times q_b \qquad (4.8)$$

where

$c_e(z)$ = exposure factor = $q_p(z)/q_b$, (4.9)
and $q_p(z)$ = peak velocity pressure at a height z in metres; q_b = basic velocity pressure
q_b = basic velocity pressure = $0.5 \times \rho \times v_b^2$, (4.10)
ρ = air density (recommended value = 1.25 kg/m³),
v_b = basic wind velocity = 24 m/s (previously assumed).

So,

$$q_b = 0.5 \times 1.25 \times v_b^2 = 0.5 \times 1.25 \times 24^2 = 360 \text{ N/m}^2.$$

Referring to Fig. 4.2 of EC 1 Part 1-4 (“Wind actions”), with a building height of $z = 11.5$ m and category II terrain, we conclude that $c_e(z) = 2.5$. Therefore, the peak velocity pressure at 11.5 m height above ground level is

$$q_p(z) = 2.5 \times 360 \text{ N/m}^2 = 0.9 \text{ kN/m}^2.$$

7.2.1.4 Wind pressure acting on external surfaces

The wind pressure acting on an external surface is given by the following expression:

$$w_e = q_p(z_e) \times c_{pe} \qquad (5.1)$$

where

c_{pe} = pressure coefficient for the external pressure, which depends on the ratio of z/d of the building,
z_e = total height of the building up to the apex with a pitched roof = 11.5 m,
d = depth of the building = 16 m (see Fig. 5.1 in Chapter 5).

Referring to Table 7.1 of EC 1: Part 1-4, with $z/d = 11.5/16 = 0.7$, or 1 say, the external pressure coefficients c_{pe} are

on windward face $c_{pe} = +0.8$;
on leeward face $c_{pe} = -0.5$.

The internal pressure coefficient (suction) on the vertical wall may be given by the following expression:

$$c_{pi} = 0.75 \times c_{pe} \qquad (7.1)$$

Thus,

on windward vertical wall, $c_{pe} = -0.75 \times 0.8 = -0.6$ (acting inwards, i.e. suction);
on leeward vertical wall, $c_{pe} = -0.75 \times 0.5 = -0.4$ (acting inwards, i.e. suction).

The resultant pressure coefficients (with internal suction) are then

on windward vertical wall, $c_{pe} = +0.8 - (-0.6) = +1.4$;
on leeward vertical wall, $c_{pe} = -0.5 - (-0.4) = -0.1$.

So the effective wind pressure on the building becomes

$$w_e = 0.9 \times [1.4 - (-0.1)] = 0.9 \times 1.5 = 1.35 \text{ kN/m}^2$$

(assumed to be acting on the external face of the building at right angles to the frames). Therefore, the characteristic wind pressure per unit height of frame with 10.5 m spacing is $w_k = 1.35 \times 10.5 = 14.2$ kN/m height.

7.2.2 Wind pressure on the gable end

The wind pressure calculations are based on EC 1. The resultant pressure on the windward gable face (previously calculated) is

$$w_e = 0.9 \times 1.4 = 1.26 \text{ kN/m}^2 \text{ (unfactored).}$$

The wind load distribution on the gable end and the forces on the vertical end bracing system are shown in Fig. 7.1a.

The uniformly distributed load on the end column/m height is given by

$$w_e \times \text{spacing (gable side see Fig. 7.1a)} = 1.26 \times 5/2 = 3.15 \text{ kN/m height.}$$

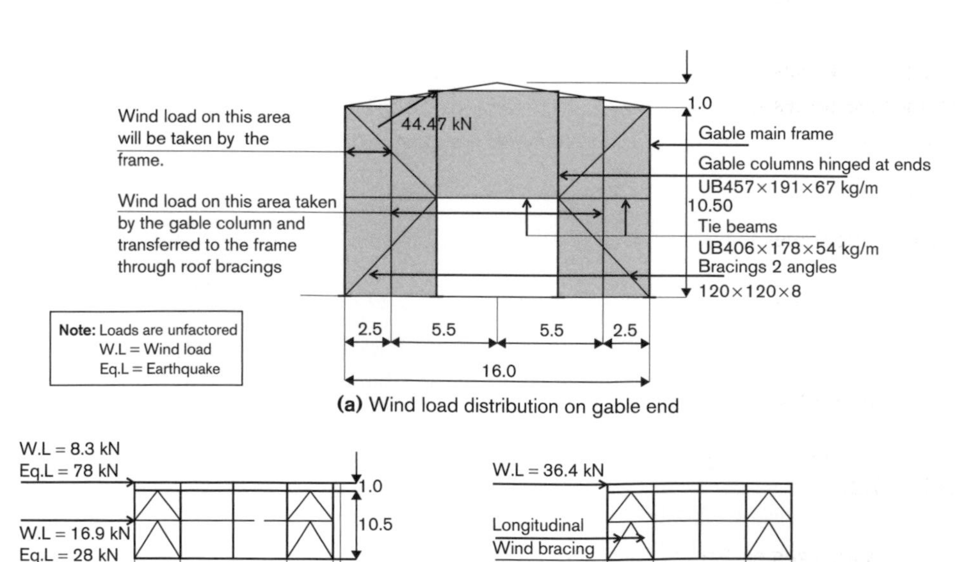

Fig. 7.1. Gable end framing system and longitudinal bracing system for seismic forces

Assuming the load is acting at the node points, the point load at eaves level is

$3.15 \times 5.25/2 = 16.9$ kN

(see Fig. 7.1b). Also, the top end reaction from the gable column which is ultimately transferred at eaves level of the end column of the end frame through the horizontal roof bracing system is

$w_e \times$ spacing of gable column $\times$ height of gable column/2 $= 1.26 \times 5.5 \times 10.5/2 = 36.4$ kN

(see Fig. 7.1c). Thus, the total wind load at eaves level of the bracing system is

$8.3 + 36.4 = 44.7$ kN

and the wind load at mid height tie beam level of the bracing system = 16.9 kN.

7.2.3 Seismic force on the bracing system

Consider the outer length of the building up to the separation joint to be 42 m. This portion of the building is provided with two sets of bracing system at each end (see Fig. 5.1). Now, these two sets of bracing system will resist the seismic forces induced on the four portal frames (see Fig. 5.1). The seismic force on each frame is 14 kN acting at mid height of the frame and also 39 kN acting at eaves level (see Figs. 8.5 and 8.6 in Chapter 8). Therefore,

- total seismic forces on each bracing system along the column line at mid height = $14/2 \times 4 = 28$ kN (unfactored) (see Fig. 7.1b);
- total seismic forces acting on each bracing system along the column line at eaves level = $39/2 \times 4 = 78$ kN (unfactored) (see Fig. 7.1b).

7.2.4 Analyses of forces in the bracing system

Consider the total wind and seismic forces to be resisted by each set of bracing system located at the ends of the building.

7.2.4.1 Analyses of forces in the members due to wind

Consider the forces in the diagonal members at mid height = F_w. The total wind forces at ground level is equal to

$$\sum H_w = (44.7 + 16.9) = 61.6 \text{ kN}.$$

Resolving the forces horizontally at the joint meeting of the two diagonals:

$$\sum H = 0: \qquad \therefore 2 \times F_w \times \cos\theta = 61.6 = 61.6.$$

So,

$$F_w = \pm 61.6/(2 \times \cos 45°) = \pm 43.6 \text{ kN}$$

(compression or tension) and the force in the horizontal member = 61.6 kN (compression) (see Fig. 7.2).

7.2.4.2 Analyses of forces in the members due to seismic forces

Consider the forces in the diagonals at mid height = F_E. Resolving the forces horizontally at the joint meeting the two diagonals:

$$\sum H = 0: \qquad \therefore 2 \times F_E \times \cos\theta = (78 + 28) = 106 \text{ kN}.$$

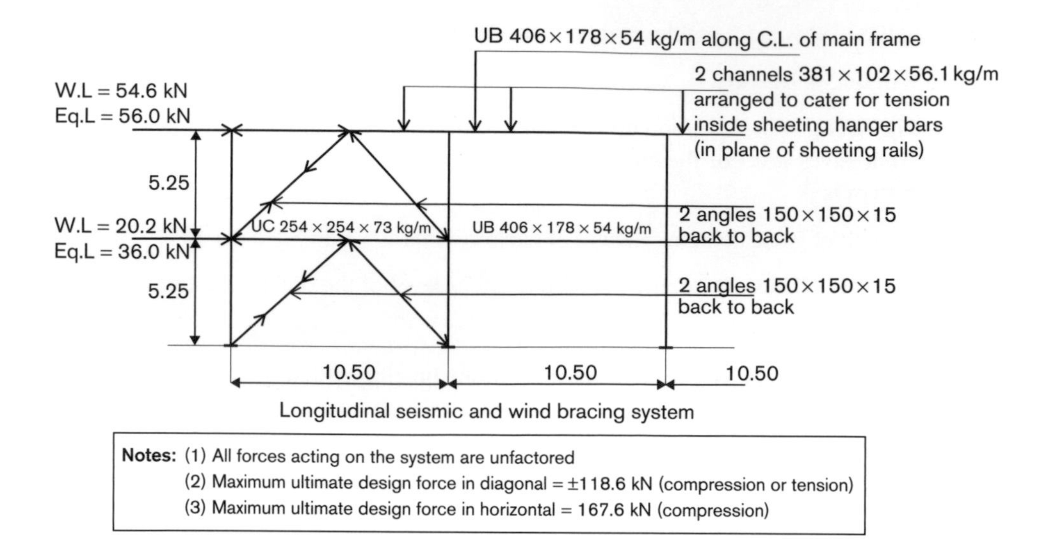

Fig.7.2. Longitudinal bracing system and member sizes

So,

$$F_E = \pm 106/(2 \times \cos 45°) = \pm 75 \text{ kN}$$

(compression or tension) and the force in the horizontal member = 106 kN (see Fig. 7.2).

7.2.5 Design of ultimate load combinations for the members

7.2.5.1 Case 1: Only wind force is acting

Referring to Table A1.2(B), "Design values of actions (STR/GEO) (Set B)", BS EN 1990: 2002(E) (Eurocode, 2002), with a partial safety factor of $\gamma_{Q,1} = 1.5$ for wind, the ultimate design force is equal to

in diagonal $N_{ed} = 1.5 \times F_w = 1.5 \times 43.6 = \pm 65.4$ kN (compression or tension);
in horizontal member = 1.5 × 61.6 = 92.4 kN (compression).

7.2.5.2 Case 2: Wind and seismic forces acting simultaneously

Referring to Table A1.3, "Design of actions for use in accidental and seismic combinations of actions", BS EN 1990: 2002(E) (Eurocode, 2002), the load combination is

$$A_{Ek} + \psi_{2,i} \times Q_k.$$

where $\psi_{2,i} = 1.0$ for the wind load assumed to be acting simultaneously as an earthquake. Therefore, the ultimate design forces are

in diagonal $N_{ed} = 1.0 \times 75 + 1.0 \times 43.6 = \pm 118.6$ kN (compression or tension);
in horizontal member = 1.0 × 106 + 1 × 61.6 = 167.6 kN (compression).

So, case 2 gives the governing design values of the members. So the maximum ultimate design forces are

in diagonal = 118.6 kN (compression or tension);
in horizontal member = 167.6 kN (compression).

7.2.6 Design of section of the members (based on EC 3)

7.2.6.1 Diagonal member

As a compression member, the maximum ultimate design force = 118.6 kN (compression). We will try two angles 150 × 150 × 10 back to back, with a 12 mm gap:

$A = 58.6\ \text{cm}^2$; $r_y = 4.62$ cm
L_y = buckling length = $L_{cr} = 5.25\ \text{m} \times \sqrt{2} = 7.42$ m (assuming bolt connections at the ends)

Referring to EC 3 (Eurocode, 2005), the design buckling resistance is

$N_{b,Rd} = \chi \times A \times f_y/\gamma_{M1}$ for class 1, 2 and cross-sections (6.47)
f_y = yield strength
γ_{M1} = resistance of members to instability assessed by member checks

where

$\chi = 1/(\Phi + (\Phi^2 - \bar{\lambda}^2)^{0.5})$ but $\chi \leq 1$, (6.49)
$\Phi = 0.5[1 + \alpha(\bar{\lambda} - 0.2) + \bar{\lambda}^2]$,
α = an imperfection factor.

The recommended value of α is given in Table 6.1 of EC 3 (see Appendix). Referring to Table 6.2 of EC 3 for a rolled L cross-section (see Appendix) and referring to the buckling curve in Fig. 6.4 of EC 3 (see Appendix), we select buckling curve "b". Then, referring to Table 6.1 of EC 3 with buckling curve "b", we obtain $\alpha = 0.34$. Using the following expression from EC 3:

$\bar{\lambda} = L_{cr}/(r_y \times \lambda_1)$ (6.50)

where

$\lambda_1 = 93.9 \times \varepsilon = 93.9 \times 0.92 = 86.4$,
$\bar{\lambda} = 7420/(4.62 \times 10 \times 86.4) = 1.86 \leq 2$ should not be greater than 2 (referring to Clause 6.7.3(3) of EC 8) (Eurocode, 2004), Satisfactory

and referring to Fig. 6.4 (of EC 3), and following the curve "b" with $\bar{\lambda}$ = 1.86 and $\chi = 0.23$, we obtain

$N_{b,Rd} = \chi \times A \times f_y/\gamma_{M1} = 0.23 \times 58.6 \times 100 \times 275/10^3 = 370\ \text{kN} > 137.6\ \text{kN}$,

with

$\Phi = 0.5[1 + 0.34 \times (1.86 - 0.2) + 1.86^2] = 2.5$;
$\chi = 1/[\Phi + (\Phi^2 - \bar{\lambda}^2)^{0.5})] = 1/[(2.5 + (2.5^2 - 1.86^2)^{0.5})] = 1/6.67 = 0.23$.

Therefore

$N_{b,Rd} = 0.23 \times 46.4 \times 100 \times 275/10^3 = 37\ \text{kN} > 293\ \text{kN}$

and

$N_{Ed}/N_{b,Rd} = 118.6/293 = 0.40 < 1$.

But for a class 3 section classification, in order to satisfy the condition

$(b + h)/2t \leq 11.5\varepsilon$

(see Table 5.2 of EC 3), where $b = h$ = depth of angle and $\varepsilon = (235/f_y)^{0.5}$, we have to increase the size of the angles. Therefore we adopt two angles 150 × 150 × 15 back to back, with 12 mm space between the angles.

7.2.6.2 Horizontal member: beam

The maximum ultimate design force = 167.6 kN (compression). Try UC254 × 254 × 73.1 kg/m, with

$L_{cr} = L_z$ = effective length = 10.5 m,
i_{rz} = 6.48 cm,
A = 93.1 cm^2,
W_z = 465 cm^3,
t_f = thickness of flange = 14.2 mm,
t_w = thickness of web = 8.6 mm,
r = root radius = 12.7 mm,
h = depth of member = 254.1 mm,
h_w = depth between fillets = 200.3 mm,
b = width = 254.6 mm,
c = outstand of flange = $[b - (t_w + 2 \times r)]/2 = [254.6 - (8.6 + 2 \times 12.7)]/2 = 110.3$ mm,
f_y = 275 N/mm^2.

- **Section classifications**

  For class 1 section classification:

  $$c/t_f \leq 9 \times \varepsilon.$$

  We determine

  $$\varepsilon = (235/f_y)^{0.5} = (235/275)^{0.5} = 0.92$$
  $$c/t_f = 110.3/14.2 = 7.78$$

  and

  $$9\varepsilon = 9 \times 0.92 = 8.28.$$

  Since $c/t_f\,(7.78) < 9\varepsilon\,(8.28)$, the condition for class 1 section classification is satisfied. Also, for class 1 section classification, the following must hold:

  $$d/t_w \leq 72\varepsilon.$$

  We find

  $$h_w/t_w = 200.3/8.6 = 23.3$$

  and

  $$72\varepsilon = 72 \times 0.92 = 66.24.$$

  Since h_w/t_w (23.3) < 72ε (66.24), the condition for class 1 section classification is satisfied.

- **Buckling resistance in compression**

  The design buckling resistance of a compression member should be taken as:

  $$N_{b,Rd} = \chi \times A \times f_y/\gamma_{M1}$$

  where

  χ = reduction factor,
  A = gross cross-section area = 93.1 cm^2,
  f_y = yield strength = 275 N/mm^2,

γ_{M1} = partial factor for resistance of the member = 1.0 (see Clause 6.1),
$\bar{\lambda}$ = slenderness for flexural buckling = $L_{cr}/(i_z \times \lambda_1)$
= 10.5 × 100/(6.48 × 86.4) = 1.88,
(where, $\lambda_1 = 93.9\varepsilon = 86.4$).

To obtain the reduction factor χ from the buckling curve in Fig. 6.4 of EC 3, go to Table 6.2 of EC 3 (see Appendix) for the selection of the buckling curve for a cross-section. For a rolled section with *h*/*b* (254.1/254.5) ≤ 1.2 and t_f(14.2 mm) ≤ 100 mm, and buckling about the *z*–*z* axis, select the buckling curve "c" in Fig. 6.4 with $\bar{\lambda} = 1.88$, $\chi = 0.2$. So

$$N_{b,Rd} = \chi \times A \times f_y/\gamma_{M1} = 0.2 \times 93.1 \times 100 \times 275/1.0/10^3$$
$$= 512 \text{ kN} > N_{ed}\ (167.6 \text{ kN})$$ Satisfactory

and

$$N_{Ed}\ (167.6 \text{ kN})/N_{b,Rd}\ (512) = 0.33 < 1.0.$$

Therefore we adopt UC 254 × 254 × 73.1 kg/m.

7.2.6.3 Design of tension members

The tension force in the diagonal members = 118.6 kN. Try two angles 150 × 150 × 10, with

$A = 58.6 \text{ cm}^2$
$N_{pl,Rd}$ = design plastic resistance of the gross cross-section
$= A \times f_y/\gamma_{M0} = 58.6 \times 100 \times 275/10^3 = 1612 \text{ kN}$
$N_{u,Rd}$ = design ultimate resistance of the net cross-section at holes
$= 0.9 \times A_{net} \times f_u/\gamma_{M2}$

where

A_{net} = A-hole area, assuming 24 mm HSG bolts,
F_u = 430 N/mm² for S 275 grade steel,
γ_{M2} = 1.25 for tension member (see Clause 6.1 of EC 3),

Net area of hole with staggered pitch

p = spacing of centres of the same two holes, measured perpendicular to the member axis = 55 mm,
s = staggered pitch, the spacing of the centres of two consecutive holes in the chain measured parallel to the member axis = 55 mm,
t = thickness of member = 10 mm,
n = number of holes extending in any line or zigzag line progressively across the member = 2,
d_0 = diameter of hole = 24 mm.

The area of deduction is

$$t[n \times d_0 - \sum s^2/4p] = 10 \times [2 \times 24 - 55^2/4 \times 55)] = 343 \text{ mm}^2$$

and

$$A_{net} = 58.6 - 3.43 \times 2 = 51.7 \text{ cm}^2$$

which means

$$N_{u,Rd} = 0.9 \times 51.7 \times 100 \times 430/(1.25 \times 1000) = 1601 \text{ kN} >> 118.6 \text{ kN}.$$

But this member may act as a compression member due to an earthquake causing a reversal of stresses. For the tension members we therefore adopt two angles 150 × 150 × 15 back to back, with 12 mm space between the angles.

7.3 Analysis and design of the column in a gable framing system

7.3.1 Design considerations

Consider a gable column 11.0 m high and spaced at 5.5 m, which is fixed at the bottom and hinged at both the base and the top. The column takes the wind and seismic forces in addition to dead loads of side sheeting and side rails.

7.3.2 Loadings

7.3.2.1 Characteristic dead loads

20G (1 mm) corrugated galvanised steel sheeting		= 0.12 kN/m^2
Insulation		= 0.028 kN/m^2
Service loads (lightings, sprinklers, etc.)		= 0.10 kN/m^2
Weight of side rails (assumed 2 m spacing)	(say)	= 0.27 kN/m^2
Self-weight of column	(say)	= 0.23 kN/m^2

	Total G_k	= 0.75 kN/m^2

Therefore the total load on the column = 0.75 × 5.5 × 11.0 = 45.4 kN.

7.3.2.2 Characteristic seismic force on the column

Design base shear force = F_{bd} = 0.37 *mg* (calculated previously).
The length of building up to the building movement joint = 42 m.
So, the design base shear force/m length = 0.37/42 = 0.009 *mg*/m length.
Spacing of column = 5.5 m.
Therefore, design base shear force on the column is

$$F_{bd} = 0.009 \times 5.5 = 0.05 \text{ kN}$$

where

m = mass of dead weight of the gable frame,
g = acceleration.

The design base shear force due to seismic action is

$$E_k = F_{bd} \times mg = 0.05 \times 45.4 = 2.27 = 3 \text{ kN, say.}$$

7.3.2.3 Characteristic wind force on the column of the gable end

The resultant wind pressure on windward gable face is

$$w_e = 0.9 \times 1.4 = 1.26 \text{ kN/m}^2 \text{ (previously calculated).}$$

The wind force on the column/m height = 1.26 × 5.5 (spacing) = 6.93 kN/m height.

7.3.3 Characteristic moment

The column is assumed to be fixed at the base and hinged at the top.

7.3.3.1 Due to seismic load

If the full base seismic force (3 kN) is considered acting at mid height of the column then

–ve. moment at base = $3 \times A_{Ek} \times L/16 = 3 \times 3.0 \times 11/16 = 6.2$ kN m;
+ve. moment at mid span = $5 \times A_{Ek} \times L/32 = 5 \times 3.0 \times 11/32 = 5.2$ kN m.

7.3.3.2 Due to wind load

−ve. moment at base = $-w_k \times L^2/8 = 6.93 \times 11^2/8 = 104.8$ kN m;
+ve. moment at mid height = $w_k \times 9 \times 11^2/128 = 59.0$ kN m.

7.3.4 Ultimate design force due to seismic action

With load combinations in ultimate limit state:

$$\gamma_1 \times A_{Ek} + \psi_2 \times W_k$$

where

γ_1 = partial factor = 1.0 (see Clause A1.3.2 of BS EN 1990: 2002(E)),
ψ_2 = combination coefficient for accompanying variable action = 0.8 (recommended for category E storage area building) (see Table A1.1 of BS EN 1990: 2002(E)),
A_{Ek} = characteristic value of seismic action,
W_k = characteristic value of wind action.

we have

ultimate design moment $M_{Ed} = 6.2 + 0.8 \times 104.8 = 90.0$ kN m;
ultimate design thrust $N_{ed} = 1.35 \times 45.4 = 61.3$ kN;
ultimate design shear at base $V_{Ed} = 3 \times 11/16$ (seismic) + $6.93 \times 11.0 \times 5/8$ (wind) = 49.7 kN.

7.3.5 Design of section

M_{Ed} = 90.0 kN m at base.
N_{Ed} = 61.3 kN.
V_{Ed} = 49.7 kN.

7.3.5.1 Initial sizing of section

Try a section UB 457 × 191 × 6/1 kg/m; steel grade S 275; f_y = 275 N/mm2.

Properties of section

Depth of section h = 453 mm.
Depth of section between fillets h_w = 407.6 mm.
Width of section b = 189.9 mm.
Thickness of web t_w = 8.5 mm.
Thickness of flange t_f = 12.7 mm.
Root radius r = 10.2 mm.
Radius of gyration i_z = 4.12 cm.
Plastic modulus W_y = 1470 cm^3.

7.3.5.2 Material

Referring to Clause 6.2 of EC 8:

- The structural steel shall confirm to standards according to BS EN 1993.
- The actual maximum yield strength $f_{y,max}$ of the steel of dissipative zones satisfies the following expression:

  $f_{y,max} \leq 1.1 \times \gamma_{ov} \times f_y$

  where

  γ_{ov} = overstrength factor (recommended value = 1.25),
  f_y = nominal yield strength specified for steel grade S 275 = 275 N/mm^2 (adopted).

Dissipative zones are expected to yield before other zones leave the elastic range during an earthquake. Thus, for grade S 275 steel,

$f_{y,max} \leq 1.1 \times 1.25 \times 275 \leq 378$ N/mm^2.

7.3.5.3 Section classification

Flange

To satisfy the condition for class 1 section classification, we must have

limiting value of $c/t_f \leq 9\varepsilon$

where

stress factor $\varepsilon = (235/275)^{0.5} = 0.92$,
outstand of flange $c = (b - t_w - 2r)/2 = (189.9 - 8.5 - 2 \times 10.2)/2 = 80.5$ mm.

The limiting value of $c/t_f = 80.5/12.7 = 6.33$ and $9\varepsilon = 9 \times 0.92 = 8.28$, so for class 1 section classification:

limiting value of c/t_f (6.33) $\leq 9\varepsilon$ (8.28). <u>Satisfactory</u>

Web

To satisfy the condition for class 1 section classification, we must have

limiting value of $h_w/t_w < 72\varepsilon$

where

ratio $h_w/t_w = 407.7/8.5 = 47.96$,
$72\varepsilon = 72 \times 0.92 = 66.3$.

Thus

limiting value of h_w/t_w (47.96) $< 72\varepsilon$ (66.3). <u>Satisfactory</u>

7.3.5.4 Moment capacity

Where the bending moment, shear and axial force act simultaneously, the moment capacity shall be calculated in the following way:

- Where shear and axial force are present, allowance should be made for the effect of both shear force and axial force on the resistance moment.

- Provided that the design value of the shear force V_{Ed} does not exceed 50% of the design plastic shear resistance, no reduction of the resistance defined for bending moment and axial force shall be made, except where shear buckling reduces the section resistance.
- Where V_{Ed} exceeds 50% of $V_{pl,Rd}$, the design resistance of the cross-section to a combination of the moment and axial force should be calculated using the yield strength $(1 - \rho) \times f_y$ of the shear area, where, $\rho = (V_{Ed}/V_{pl,Rd} - 1)^2$.

(a) *When the web is not susceptible to buckling.*

When the web depth to thickness ratio $h_w/t_w \leq 72\varepsilon$ for class 1 section classification, it should be assumed that the web is not susceptible to buckling, and the moment capacity shall be calculated by the expression $M_{y,Rd} = f_y \times W_y$, provided the shear force $V_{Ed} < V_{pl,Rd}$. In our case, we have assumed a section in which $h_w/t_w < 72\varepsilon$. So, the web is not susceptible to buckling.

(b) *When the ultimate shear force* $V_{Ed} \leq 0.5 \times V_{pl,Rd}$.

We have

$$V_{pl,Rd} = A_v \times (f_y/\sqrt{3})/\gamma_{Mo}$$

where

A_v = shear web area = $h \times t_w = 453 \times 8.5 = 3851$ mm^2,

γ_{Mo} = partial factor = 1.0.

Therefore, in our case, $V_{pl,Rd} = 3851 \times (275/1.73)/10^3 = 612$ kN, and $0.5 \times V_{pl,Rd} = 612/2 = 306$ kN. Thus,

$$V_{Ed}\ (49.7\ \text{kN}) < 0.5 \times V_{pl,Rd}\ (306\ \text{kN})$$

so no reduction of plastic resistance moment needs to be considered.

(c) *When the ultimate axial force* $N_{Ed} \leq 0.25 \times N_{pl,Rd}(0.25 \times A \times f_y/\gamma_{Mo})$.

But, $(0.25 \times A \times f_y/\gamma_{Mo}) = 0.25 \times 85.5 \times 100 \times 275/10^3 = 588$ kN.

So, $N_{Ed}(61.3\ \text{kN}) < 0.25 \times N_{pl,Rd}(588\ \text{kN})$. Therefore, no reduction of plastic resistance moment needs to be considered and the plastic moment capacity is

$$M_{pl,Rd} = f_y \times W_{pl,y} = 275 \times 1470 \times 10^3/10^6$$
$$= 404\ \text{kN m} > M_{Ed}\ (90\ \text{kN m}).$$

Satisfactory

7.3.5.5 Shear buckling resistance

Shear buckling resistance need not be checked if the ratio $h_w/t_w \leq 72\varepsilon$. In our case, $h_w/t_w = 47.96$ and $72\varepsilon = 72 \times 0.92 = 66.24$. So, $h_w/t_w < 72\varepsilon$. Therefore, shear buckling resistance need not be considered.

7.3.5.6 Buckling resistance to compression

The buckling resistance to compression is equal to

$$N_{b,Rd} = \chi \times A \times f_y/\gamma_{M1}$$

where

χ = reduction factor,
f_y = yield strength
γ_{M1} = resistence of members to instability assessed by member checks
$\bar{\lambda} = L_{cr}/(i_z \times \lambda_1)$,

$\lambda_1 = 93.9\varepsilon = 93.9 \times 0.92 = 86.4$,
L_{cr} = buckling length in the buckling plane = 2.0 m (spacing of side rails),
$i_z = 4.12$ cm,
$\bar{\lambda} = 200/(4.12 \times 86.4) = 0.56$.

To obtain the value of χ from the buckling curve in Fig 6.4 (of EC 3), select the buckling curve from Table 6.2 (of EC 3). With a rolled section $h/b = 453.4/189.9 = 2.4 > 1.2$ and $t_f \leq 40$ mm with buckling about z–z, follow the curve "b". With $\bar{\lambda} = 0.56$ and $\chi = 0.90$, we see that

$$N_{b,Rd} = \chi \times A \times f_y/\gamma_{M1} = 0.90 \times 85.5 \times 100 \times 275/10^3$$
$$= 2116 \text{ kN} > N_{Ed}\ (61.3).$$ Satisfactory

7.3.5.7 Buckling resistance moment

The column is assumed to be fixed at the base and hinged at the top. The outer compression flange in contact with the side rails spaced at 2 m intervals is restrained against horizontal buckling. But the inner compression flange is not restrained against buckling for about a quarter of the height (2.75 m) from the base at the point of contraflexure in the bending moment diagram. So, this part of the compression flange is unrestrained, and the buckling resistance moment shall be calculated in the following way. Referring to Clause 6.3.2.4 of EC 3 ("Simplified assessment methods for beams with restraints in buildings"), members with lateral restraint to the compression flange are not susceptible to lateral torsional buckling if, given the length L_c between restraints, the resulting slenderness $\bar{\lambda}_f$ of the equivalent compression flange satisfies the following expression:

$$\bar{\lambda}_f = k_c \times L_c/(i_{f,z} \times \lambda_1) \leq \bar{\lambda}_{c0} \times (M_{c,Rd}/M_{y,Ed})$$

where

$M_{c,Rd} = W_y \times f_y/\gamma_{M1} = 1470 \times 10^3 \times 275/10^6 = 404$ kN m,
$M_{y,Ed}$ = maximum design moment value within the restraint spacing = 90.0 kN m,
W_y = modulus of section corresponding to the compression flange = 1470 cm^3,
k_c = slenderness correction factor = 0.91 (see Table 6.6 of EC 3),
$\bar{\lambda}_{c0}$ = slenderness limit of the equivalent compression flange = $\bar{\lambda}_{LT,0} + 0.1 = 0.4 + 0.1 = 0.5$ (see Clause 6.3.2.3 of EC 3),
$\lambda_1 = 93.9 \times \varepsilon = 86.4$,
$i_{f,z}$ = radius of gyration of the equivalent compression flange about the minor axis = 4.12 cm,
L_c = 275 cm.

Therefore

$$\bar{\lambda}_f = k_c \times L_c/(i_{f,z} \times \lambda_1) = 0.91 \times 275/(4.12 \times 86.4) = 0.70$$

and

$$\bar{\lambda}_{c0} \times (M_{c,Rd}/M_{y,Ed}) = 0.5 \times (404/90) = 2.24.$$

Thus

$$k_c \times L_c/(i_{f,z} \times \lambda_1)\ (0.70) < \bar{\lambda}_{c0} \times (M_{c,Rd}/M_{y,Ed})\ (2.24)$$

So, reduction of the design buckling resistance moment is unnecessary. Also

$$M_{y,Ed}/M_{c,Rd} + N_{Ed}/M_{b,Rd} = 90/404 + 61.3/2116 = 0.22 + 0.03 = 0.15 < 1.0.$$

Satisfactory

Therefore, adopt UB 457 × 191 × 6/1 kg/m.

References

Eurocode, 2002. BS EN 1990: 2002, Basis of Structural Design.
Eurocode, 2004. BS EN 1998-1: 2004, Eurocode 8. Design of Structures for Earthquake Resistance.
Eurocode, 2005. BS EN 1993-1-1: 2005, Eurocode 3, Design of Steel Structures. Eurocode, 2005. BS EN 1991-1-4: 2005 (E), Eurocode 1: Part 1-4 - Wind actions.

CHAPTER 8

Case Study: Analysis of a Portal Frame With Seismic Forces

8.1 Loadings

See Figs. 5.1, 5.2 and 5.3 in Chapter 5 for plan and section views.

8.1.1 Characteristic dead loads (G_k) and imposed loads, based on EC 1, Part 1-1 (Eurocode, 2002a)

(1) Roof

20G (1 mm) corrugated galvanised steel sheeting	= 0.12 kN/m^2
Insulation	= 0.028 kN/m^2
Service loads (lightings, sprinklers, etc.)	= 0.10 kN/m^2
Weight of purlins (assumed UB 406 × 178 × 5/1 kg/m) (assumed purlin spacing = 2 m)	= 0.27 kN/m^2
Self-weight of frame (assumed UB 610 × 305 × 238 kg/m) = 238/10.5/1000 (assumed frame spacing = 10.5 m)	= 0.23 kN/m^2
Total	= 0.75 kN/m^2

∴ dead loads/m of roof beam = 0.75 × 10.5 = 7.85 kN/m

(2) Side sheeting

20G (1 mm) corrugated galvanised steel sheeting	= 0.12 kN/m^2
Insulation	= 0.028 kN/m^2
Weight of sheeting rail (assumed UB 406 × 178 × 5/1 kg/m)	= 0.27 kN/m^2
Self-weight of column (assumed UB 610 × 305 × 23/1 kg/m)	= 0.23 kN/m^2
Total	= 0.65 kN/m^2

∴ dead loads/m of column = 0.65 × 10.5 = 6.83 kN/m height

8.1.2 Characteristics live loads (Q_k)

The live loads/m of the roof with access = 1.5 kN/m^2 × 10.5 (spacing) = 15.75 kN/m.

8.1.3 Characteristic wind loads (W_k), based on EC 1, Part 1-4 (Eurocode, 2006)

8.1.3.1 Wind pressure acting on external surfaces

The wind pressure acting on an external surface is given by the following expression:

$$w_e = q_p(z_e) \times c_{pe}$$

where

q_p = peak velocity pressure = 0.9 kN/m^2 (previously calculated in Chapter 7),
c_{pe} = pressure coefficient for the external pressure, which depends on the ratio of z/d of the building,
z_e = total height of the building up to the apex with a pitched roof = 11.5 m,
d = depth of the building = 16 m (see Fig. 5.1 in Chapter 5).

Referring to Table 7.1 of EC 1, Part 1-4 (Eurocode, 2006), with $z/d = 11.5/16 = 0.7$ (= 1, say), we find the external pressure coefficients are

on windward face $c_{pe} = +0.8$;
on leeward face $c_{pe} = -0.5$.

The internal pressure coefficient (suction) on the vertical wall may be given by the following expression:

$$c_{pi} = 0.75 \times c_{pe}. \tag{7.1}$$

Therefore

on windward vertical wall $c_{pe} = -0.75 \times 0.8 = -0.6$ (acting inwards, i.e. suction);
on leeward vertical wall $c_{pe} = -0.75 \times 0.5 = -0.4$ (acting inwards, i.e. suction).

The resultant pressure coefficients (with internal suction) are

on windward vertical wall = $+0.8 - (-0.6) = +1.4$;
on leeward vertical wall = $-0.5 - (-0.4) = -0.1$.

So, the effective wind pressure on the building is

$$w_e = 0.9 \times [1.4 - (-0.1)] = 0.9 \times 1.5 = 1.35 \text{ kN/m}^2$$

(assumed to be acting on the external face of the building). Therefore, the characteristic wind pressure/m height of frame with 10.5 m spacing is

$$w_k = 1.35 \times 10.5 = 14.2 \text{ kN/m height.}$$

8.1.4 Loadings from the overhead conveyor system

The conveyor trusses support the overhead system, and they are hung from the main frame with hanger bars. Assuming

span of conveyor truss = 10.5 m = spacing of main frame,
spacing of conveyor truss = 3 m, 1.5 m on either side of the apex,

let's consider the loads on each hanger bar connected to the roof beam of the frame.

8.1.4.1 Characteristic dead load (DL) from the overhead conveyor system (G_k) Assuming the dead load of each conveyor truss = 1.4 kN/m, the dead load on each hanger = 1.4 × 10.5 = 14.7 kN.

8.1.4.2 Characteristic live load (LL) from the overhead conveyor system Consider the conveyor and the material carried by it as moving loads, which means these loads shall be considered as live loads. Thus,

Weight of conveyor	= 0.5 kN/m run (assumed)
Weight of material in the conveyor	= 1.75 kN/m run (assumed)
Live load on the walking platform on either side	= 1 kN/m run (assumed)

Total	= 3.25 kN/m

∴ live load on each hanger = 3.25 × 10.5 = 34.13 kN

There are two hangers spaced at 3.0 m connected to the main frame. Therefore, two vertical dead loads, each 14.7 kN, and two vertical live loads, each 34.13 kN, are applied to the main frame, 1.5 m apart from either side of the apex of the frame. These dead and live loads are very close to the apex. So, consider these two dead and live loads acting as one concentrated dead and live load applied at the apex. Thus,

total characteristic dead load at apex $G_k = 2 \times 4.7 = 29.4$ kN;
total characteristic live load at apex $Q_k = 2 \times 34.13 = 68.3$ kN.

8.1.5. Seismic forces on the frame (E_k)

The design base shear force = $F_{bd} = 0.37\ mg$ (previously calculated in Chapter 6), and the building length up to the movement joint of the building = 42 m. So, the design base shear force/m length = 0.37 mg/42 = 0.009 mg/m length.

With 10.5 m spacing of steel portal frame, the design base shear force/frame $F_{bd} = 0.009 \times 10.5 = 0.095\ mg$, where m = mass of dead weight on the frame including the roof, side sheeting and partial live loads. Taking into account combinations of seismic action with other actions (dead and live loads), the design base shear force due to seismic action is

$$F_{bd} = 0.095 \times (\Sigma G_{k,i} + 0.8\Sigma Q_{k,i})\ \text{kN}$$

(see Chapter 6).

(1) Total load on the roof beam of the frame (10.5 m spacing):

From previous calculations, dead weight of roof = 7.85 × 16	= 125.6 kN
Dead weight at the apex from the conveyor	= 29.4 kN

Total dead load	= 153.4 kN
Live load on roof = 15.75 × 16	= 252 kN
Live load at the apex from the conveyor	= 68.3 kN

Total live load	= 320.3 kN

∴ design base shear force due to seismic action is

$$E_{k1} = 0.095 \times (153.4 + 0.8 \times 320.3) = 39\ \text{kN}$$

assumed to act at the eaves level of the frame.

(2) Total load on the columns of the frame (10.5 m spacing) from the side sheetings:
From previous calculations:
Dead load/m height on the column = 6.83 kN
Total dead load for full height of columns = 6.83 × 10.5 = 71.7 kN
∴ the seismic force on the column self-weight is
$E_{k2} = 0.095 \times 71.7 = 6.8$ (7 kN, say)
acting at mid height of each column, i.e. 5.25 m from the base.

8.2 Analysis of the frame

The analysis will be carried out manually with the aid of standard established formulae from the handbook *Rahmen Formeln* by Kleinlogel (1958).

- **Preparation of the structural model**
  Assume the bases are hinged and the top ends of the columns are welded to the roof beam ends to form a rigid portal frame.
- **Assumed member sizes**
  I_c = moment of inertia of the column member.
  I_b = moment of inertia of the roof beam.
  L = span (centre-to-centre distance of column) = 16 m.
  h = height from the base to eaves level = 10.5 m.
  f = height from the eaves to the apex = 1.0 m.
  s = length of the rafter = $(f^2 + (h/2)^2)^{0.5}$ = 8.06 m.
- **Coefficients**
  $k = I_b/I_c \times h/s = 10.5/8.06 = 1.3$ (assumed $I_b = I_c$).
  $\varphi = f/h = 1/10.5 = 0.095$.
  $m = 1 + \varphi = 1.095$.
  $B = 2 \times (k + 1) + m = 2 \times 2.3 + 1.095 = 5.695$.
  $C = 1 + 2 \times \mathrm{m} = 1 + 2 \times 1.095 = 3.19$.
  $N = B + m \times C = 5.695 + 1.095 \times 3.19 = 9.19$.

8.2.1 Characteristic moments, shears and thrusts

8.2.1.1 Due to characteristic dead loads (G_k = 7.85 kN/m)

$M_{ba} = M_{de}$, moment at eaves level $= -G_k \times L^2 \times (3 + 5\,m)/(16\,N)$
$= -7.85 \times 16^2 \times (3 + 5 \times 1.095)/(16 \times 9.19) = -115.8$ kN m.
$M_{cb} = M_{cd}$, moment at apex $= G_k \times L^2/8 + m \times M_{ba}$
$= 5.52 \times 16^2/8 + 1.095 \times -81.4 = 124.4$ kN m.
H_a = horizontal thrust at base A $= -\mathrm{M}_{ba}/\mathrm{h} = -(-115.8)/10.5 = 11.03$ kN.
H_e = horizontal thrust at base E = −11.03 kN.
V_a = axial thrust at base A $= G_k \times L/2 = 7.85 \times 16/2 = 62.8$ kN.
V_e = axial thrust at base E = 62.8 kN.
(See Fig. 8.1.)

8.2.1.2 Due to characteristic live loads (Q_k = 15.75 kN/m)

$M_{ba} = M_{de} = -115.8 \times 15.75/7.85 = -231.6$ kN m.
$M_{cb} = M_{cd} = Q_k \times 16^2/8 + 1.095 \times 231.6 = 250.4$ kN m.
$H_a = 11.03 \times 15.75/7.85 = 22.06$ kN.

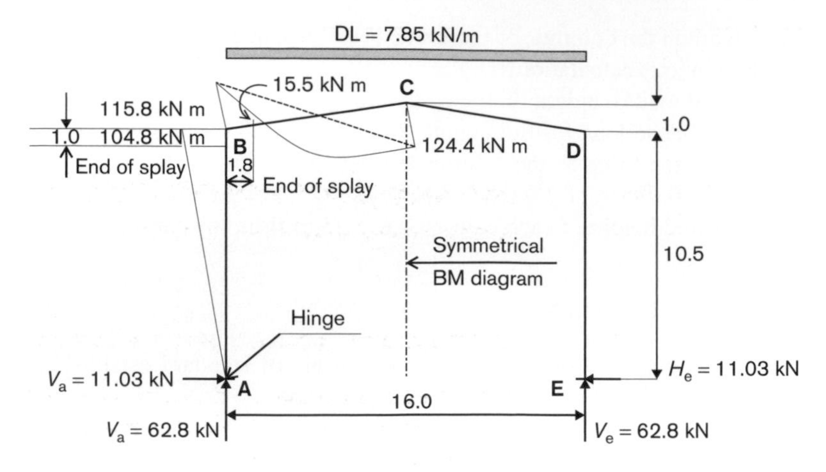

Fig. 8.1. Bending moment (BM) diagram for dead loads

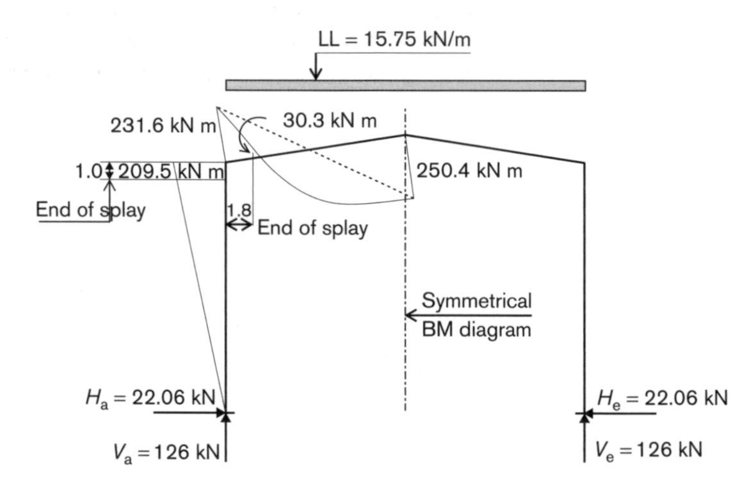

Fig. 8.2. BM diagram for live loads

$H_e = -22.06$ kN.
$V_a = Q_k \times L/2 = 15.75 \times 16/2 = 126$ kN.
$V_e = 126$ kN.
(See Fig. 8.2.)

8.2.1.3 Due to characteristic wind loads (W_k = 14.2 kN/m height on the column)

The wind is assumed to be blowing left to right and up to eaves level.

$M_{de} = -W_k \times h^2/8 \times [2 \times (B + C) + k]/N$
$= -14.2 \times 10.5^2/8 \times [2 \times (5.695 + 3.19) + 1.3]/9.19 = -406$ kN m.
$M_{ba} = W_k \times h^2/2 + M_{de} = 14.2 \times 10.5^2/2 - 406 = 376.8$ kN m.
$M_{cb} = W_k \times h^2/4 + m \times M_{de}$
$= 14.2 \times 10.5^2/4 + 1.095 \times (-406) = -53.2$ kN m.

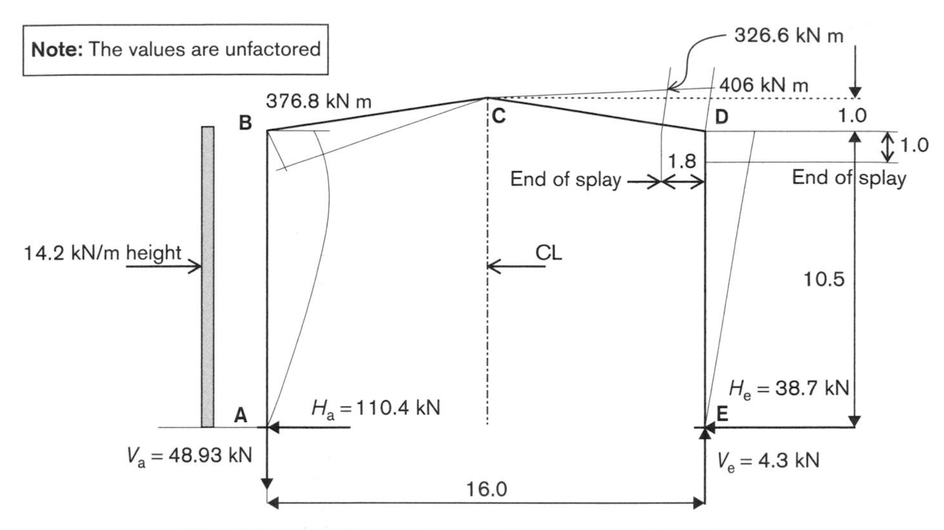

Fig. 8.3. BM diagram for wind loads on the left columns

Taking the moment at the right-hand base hinge,

$V_a = -W_k \times h^2/2L = -14.2 \times 10.5^2/(2 \times 16) = -48.9$ kN.
$V_e = +\ 48.9$ kN.
$H_e = -M_{de}/h = -406/10.5 = -38.7$ kN.
$H_a = -(W_k \times h - H_e) = -(14.2 \times 10.5 - (38.7)) = -110.4$ kN.
(See Fig. 8.3.)

8.2.1.4 Due to characteristic wind loads (W_k = 14.2 kN/m) from eaves level to the apex

Constants

$X\ = W_k \times f^2 \times (C + m)/8\,N = 14.2 \times 1^2 \times (3.19 + 1.095)/(8 \times 9.19) = 0.83.$
$M_{ba} = +X + W_k \times f \times h/2 = 0.83 + 14.2 \times 1 \times 10.5/2 = 75.4$ kN m.
$M_{de} = X - W_k \times f \times h/2 = 0.83 - 14.2 \times 10.5/2 = -73.7$ kN m.
$M_c\ = -W_k \times f^2/4 + m \times X = -14.2 \times 1/4 + 1.095 \times 0.83 = -2.6$ kN m.
$V_a\ = -W_k \times f \times h \times (1 + m)/2L = -14.2 \times 10.5 \times (1 + 1.095)/(2 \times 16) = -9.8$ kN.
$V_e\ = -(V_a) = +\ 9.8$ kN.
$H_a\ = -X/h - W_k \times f/2 = -0.83/10.5 - 14.2/2 = -7.2$ kN.
$H_e\ = -X/h + W_k \times f/2 = -0.83/10.5 + 14.2/2 = -7.0$ kN.
(See Fig. 8.4.)

8.2.1.5 Due to seismic loads at mid height of the column (E_{k2} = 7 kN)

Seismic force acting at mid height of column AB = E_{k2} = 7 kN.
Seismic force acting at mid height of column DE = E_{k2} = 7 kN.
Height from base to mid height = $a = h/2 = 10.5/2 = 5.25$ m.
Height from mid height to eaves level = $b = h/2 = 5.25$ m.

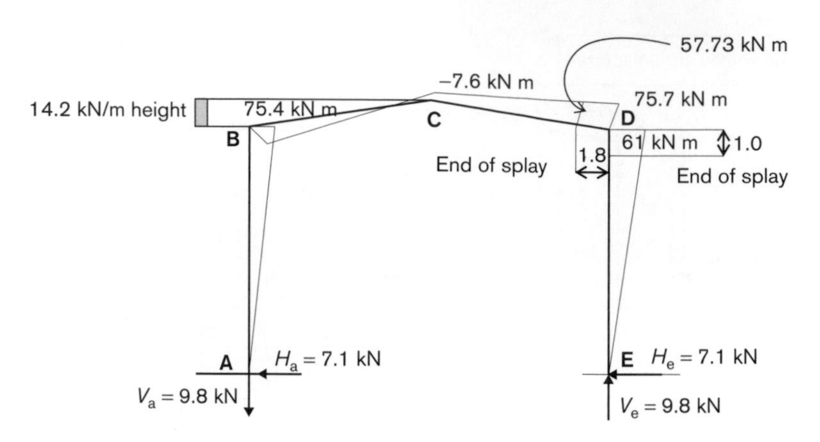

Fig. 8.4. BM diagram for wind loads from eaves level to the apex

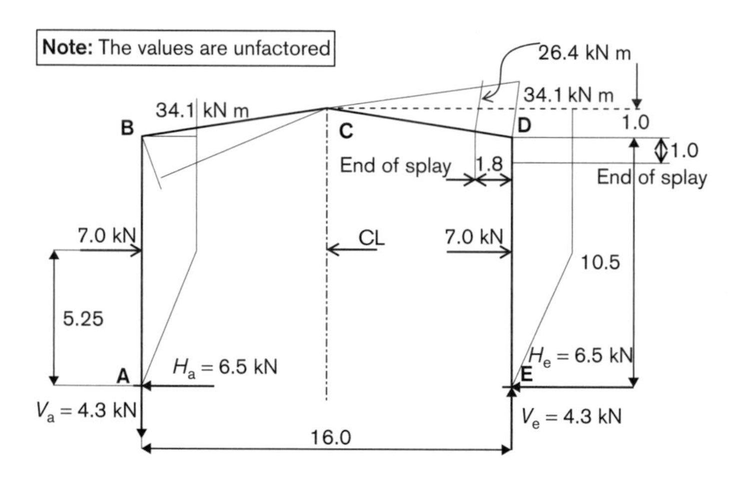

Fig. 8.5. BM diagram for seismic loads at mid height of the columns

$M_{ba} = E_{k2} \times a = 7 \times 5.25 = 36.8$ kN m; $M_{de} = -36.8$ kN m.
$M_c = 0$; moment at load points $M_x = 7 \times 5.25 = 36.8$ kN m.
$H_a = -E_{k2} = -7$ kN $= H_e$.
$V_a = -2 \times E_{k2}/2 \times a/L = -2 \times 7 \times 5.25/16 = -4.6$ kN $= V_e$.
(See Fig. 8.5.)

8.2.1.6 Due to seismic loads at eaves level (E_{k1} = 39 kN)

$M_{de} = -E_{k1} \times h \times (B + C)/2N$
$= -39 \times 10.5 \times (5.695 + 3.19)/(2 \times 9.19) = -198$ kN m.
$M_{ba} = E_{k1} \times h + M_{de} = 39 \times 10.5 - 198 = 211.5$ kN m.
$M_c = E_{k1} \times h/2 + m \times M_{de} = 39 \times 10.5/2 + 1.095 \times (-198) = -12$ kN m.
$V_a = -E_{k1} \times h/L = -39 \times 10.5/16 = -25.6$ kN $= -V_e$.
$H_e = -M_{de}/h = -198/10.5 = -18.9$ kN.
$H_a = -(E_{k1} - H_e) = -(39 - (18.9)) = -20.1$ kN.
(See Fig. 8.6.)

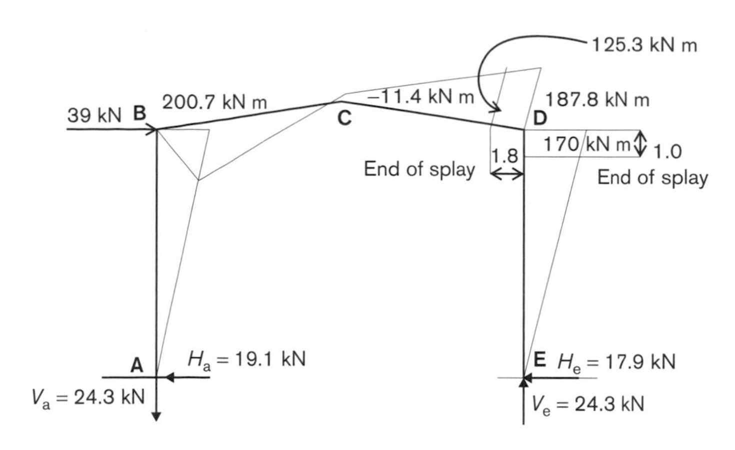

Fig. 8.6. BM diagram for seismic loads at eaves level

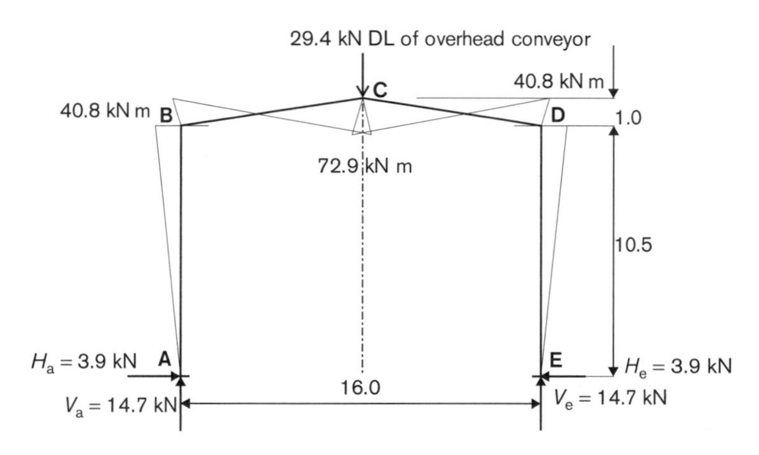

Fig. 8.7. BM diagram for dead loads of overhead conveyor system

8.2.1.7 Due to DL (point) from overhead conveyor at the apex (P_d = 29.4 kN)

$M_{ba} = M_{de} = -P_d \times L/4 \times C/N = -29.4 \times 16/4 \times 3.19/9.19 = -40.8$ kN m.
$M_c = P_d \times L/4 \times B/N = 29.4 \times 16/4 \times 5.695/9.19 = 72.9$ kN m.
$V_a = V_e = 29.4/2 = 14.7$ kN.
$H_a = M_{ba}/h = 40.8/10.5 = 3.9$ kN; $H_e = -H_a = -3.9$ kN.
(See Fig. 8.7.)

8.2.1.8 Due to LL (point) from overhead conveyor at the apex (P_l = 68.3 kN)

$M_{ba} = M_{de} = P_l \times L/4 \times C/N = -40.8 \times 68.3/29.4 = -94.8$ kN m.
$M_c = +72.9 \times 68.3/29.4 = 169.3$ kN m.
$V_a = V_e = 68.3/2 = 34.2$ kN.
$H_a = M_{ba}/h = 94.8/10.5 = 9.0$ kN; $H_e = -9.0$ kN.
(See Fig. 8.8.)

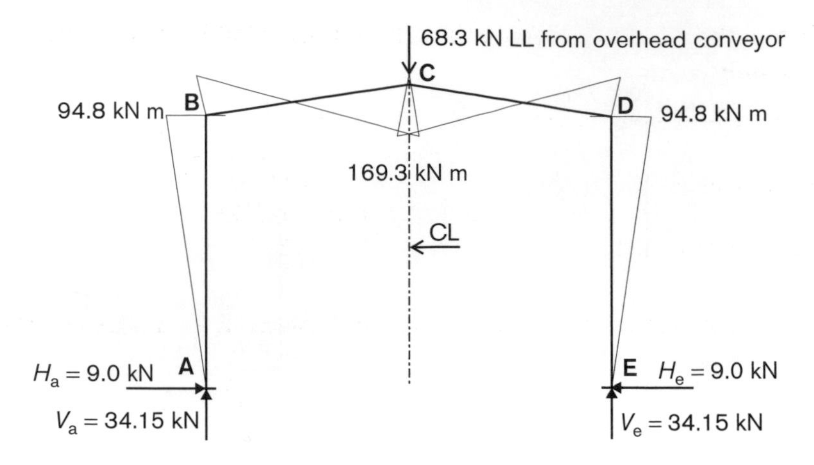

Fig. 8.8. BM diagram for live loads of overhead conveyor system

8.3 Design of structural members

A moment-resisting frame shall be designed in accordance with EC 3, Part 1-1 (Eurocode, 2005) in conjunction with EC 8 (Eurocode, 2004), so that plastic hinges form in the beams or in the connections of the beams to the columns, but not in the columns.

The specific rules for the design of steel-framed buildings of moment-resisting frames have been explained in Chapter 2.

8.3.1 Effective characteristic moments, shear and thrust in the roof member

The roof member is splayed at the connection with the column. The design section shall be taken at the end of splay (1.8 m from the centre line of the column) of the roof member, and the moments at that section will be calculated as follows.

8.3.1.1 When the dead loads cover the full span

Characteristic moment at the end of splay: $M_{gk} = V_a \times 1.8 - G_k \times 1.8^2/2 - M_{de} = 62.8 \times 1.8 - 7.85 \times 1.8^2/2 - 115.8$
$= -15.5$ kN m.

Characteristic moment at the apex $= +\,124.4$ kN m.

Characteristic shear at the end of splay: $V_{gk} = V_a - G_k \times 1.8 = 62.8 - 7.85 \times 1.8$
$= 48.7$ kN.

Characteristic thrust at the end of splay: $N_{gk} = H_a = 11.3$ kN.

8.3.1.2 When the live loads cover the full span

Characteristic moment at the end of splay: $M_{qk} = V_a \times 1.8 - Q_k \times 1.8^2/2 - 231.6$
$= 126 \times 1.8 - 15.75 \times 1.8^2/2 - 231.6$
$= -30.3$ kN m.

Characteristic moment at the apex $= +\,250.4$ kN m.

Characteristic shear at the end of splay: $V_{qk} = V_a - Q_k \times 1.8 = 126 - 15.75 \times 1.8$
$= 97.7$ kN.

Characteristic thrust at the end of splay: $N_{qk} = H_a = 22.1$ kN.

8.3.1.3 When the wind blows from the left and the wind load is on the left column only

Characteristic moment at the end of splay: $M_{wk} = M_{cd} + (M_{dc} - M_{cd}) \times 6.2/8 = 53.2 + (406 - 53.2) \times 6.2/8 = -326.6$ kN m.
Characteristic moment at the apex $= -53.2$ kN m.
Characteristic shear at the end of splay: $V_{wk} = V_e = 48.9$ kN.
Characteristic thrust at the end of splay: $N_{wk} = H_e = 38.7$ kN.

8.3.1.4 When the wind blows from the left and the wind load is on the left roof projection only

Characteristic moment at the end of splay: $M_{wk} = -2.6 - (73.7 - 2.6) \times 6.2/8 = -57.7$ kN m.
Characteristic moment at the apex $= -2.6$ kN m.
Characteristic shear at the end of splay: $V_{wk} = V_e = 9.8$ kN.
Characteristic thrust at the end of splay: $N_{wk} = N_e = 7.2$ kN.

8.3.1.5 When the seismic point load acts at mid height of the left and right columns from left to right

Characteristic moment at the end of splay: $M_{ek} = 36.8 \times 6.2/8 = 28.5$ kN m.
Characteristic moment at the apex $= 0$.
Characteristic shear at the end of splay: $V_{ek} = V_c = 4.6$ kN.
Characteristic thrust at the end of splay: $N_{ek} = 7$ kN.

8.3.1.6 When the seismic point load acts at the left eaves level

Characteristic moment at the end of splay: $M_{ek} = -12 + (198 - 12) \times 6.2/8 = -156.2$ kN m.
Characteristic moment at the apex $= -12$ kN m.
Characteristic shear at the end of splay: $V_{ek} = V_e = 25.6$ kN.
Characteristic thrust at the end of splay: $N_{ek} = H_e = 18.9$ kN.

8.3.1.7 When the overhead conveyor is running, dead load (Crdl) = 29.4 kN at the apex

Characteristic DL moment at the end of splay: $M_{cnd} = 40.8 \times 1.1/2.9 = 15.5$ kN m.
Characteristic DL moment at the apex $= 72.6$ kN m.
Characteristic shear at the end of splay: $V_{cnd} = 14.7$ kN.
Characteristic thrust at the end of splay: $N_{cnd} = 3.9$ kN.

8.3.1.8 When the overhead conveyor is running, live load (Crl) = 68.3 kN at the apex

Characteristic LL moment at the end of splay: $M_{cnl} = 94.8 \times 1.1/2.9 = 36$ kN m.
Characteristic LL moment at the end apex $= 169.3$ kN m (see Fig. 8.8).
Characteristic shear at the end of splay: $V_{cnl} = 34.2$ kN.
Characteristic thrust at the end of splay: $N_{cnl} = 9$ kN.

8.3.2 Effective characteristic moments, shear and thrust in the column

The effective characteristic moments for the design of members shall be calculated at the end of splay from the centre line of the frame as follows.

8.3.2.1 When the dead loads on the roof cover the whole span

Effective characteristic moments at the end of splay: $M_{gk1} = 115.8 \times 9.5/10.5 = 104.8$ kN m.
Effective characteristic shear at the end of splay: $V_{gk1} = 11.3$ kN.
Effective characteristic thrust at the end of spay: $N_{gk1} = 62.8$ kN.

8.3.2.2 When the live loads on the roof cover the whole span

Effective characteristic moments at the end of splay: $M_{qk1} = 231.6 \times 9.5/10.5 = 209.5$ kN m.
Effective characteristic shear at the end of splay: $V_{qk1} = 22.1$ kN.
Effective characteristic thrust at the end of splay: $N_{qk1} = 126$ kN.

8.3.2.3 When the wind blows on the left column from the left

Effective characteristic moment at the end of splay: $M_{wk1} = 406 \times 9.5/10.5 = 367.3$ kN m.
Effective characteristic shear at the end of splay: $V_{wk1} = 38.7$ kN.
Effective characteristic thrust at the end of splay: $N_{wk1} = 48.9$ kN.

8.3.2.4 When the wind blows from the left, covering the eaves to apex height only

Effective characteristic moment at the end of splay: $M_{wk2} = 73.7 \times 9.5/10.5 = 66.7$ kN m.
Effective characteristic shear at the end of splay: $V_{qk2} = 7.2$ kN.
Effective characteristic thrust at the end of splay: $N_{qk2} = 9.8$ kN.

8.3.2.5 When the seismic forces act at mid height of the column

Effective characteristic moment at the end of splay: $M_{ek1} = 36.81$ kN m.
Effective characteristic shear at the end of splay: $V_{ek1} = 7$ kN.
Effective characteristic thrust at the end of splay: $N_{ek1} = 4.6$ kN.

8.3.2.6 When the seismic forces act at eaves level

Effective characteristic moment at the end of splay: $M_{ek2} = 19.8 \times 9.5/10.5 = 179$ kN m.
Effective characteristic shear at the end of splay: $V_{ek2} = 18.9$ kN.
Effective characteristic thrust at the end of splay: $N_{ek2} = 25.6$ kN.

8.3.2.7 When the conveyor DL acts at the apex

Effective characteristic moment at the end of splay: $M_{cnk1} = 40.8 \times 9.5/10.5 = 36.9$ kN m.

Effective characteristic shear at the end of splay: V_{cnk1} = 3.9 kN.
Effective characteristic thrust at the end of splay: N_{cnk1} = 14.7 kN.

8.3.2.8 When the conveyor LL acts at the apex

Effective characteristic moment at the end of splay: M_{cnk2} = 94.8 × 9.5/10.5 = 85.8 kN m.
Effective characteristic shear at the end of splay: V_{cnk2} = 9 kN.
Effective characteristic thrust at the end of splay: N_{cnk2} = 34.2 kN.

8.4 Design load combinations (ULS method)

Case 1: Combinations of actions for seismic design situations.
Referring to Clause 6.4.3.4, "Combinations of actions for seismic design situation", of BS EN 1990: 2002(E) (Eurocode, 2002b), the load combinations are given by the following expression:

$$\Sigma\gamma_{Gj} \times G_{k,j} + A_{Ed} + \Sigma\psi_{2,i} \times Q_{K,i} \quad (6.12b)$$

where

$G_{k,j}$ = characteristic value of permanent action of dead loads,
γ_{Gj} = partial factor = 1.0 for seismic load combinations,
$Q_{K,i}$ = variable loads,
A_{Ed} = design value of leading seismic action = $\gamma_1 \times A_{Ek}$,
A_{Ek} = characteristic value of seismic action,
γ_1 = 1.0 (refer to Clause A1.3.2 of BS EN 1990: 2002(E) "Design values of actions in the accidental and seismic design situations"),
$\psi_{2,i}$ = combination coefficient for accompanying variable action main
= 0.8 (recommended for category E storage area building – see Table A1.1 of BS EN 1990: 2002(E) for recommended values of ψ factors for buildings),
$Q_{k,i}$ = accompanying variable actions.

Case 2: Combinations of actions without seismic design situations.
Referring to Clause 6.4.3.2, "Combinations of actions for persistent or transient design situation (fundamental combinations)", of EN 1990: 2002 (E) (Eurocode, 2002b), the load combinations are given by the following expression:

$$\Sigma\gamma_{G,j} \times G_{k,j} + \gamma_{Q,1} + \Sigma\gamma_{Q,i} \times \psi_{0,i} \times Q_{k,i} \quad (6.10)$$

where

$\gamma_{G,j}$ = partial factor for permanent action = 1.35,
$G_{k,j}$ = characteristic value of permanent action (DL),
$\gamma_{Q,1}$ = partial factor for leading variable action = 1.5 (see Table A1.2(A) of BS EN 1990: 2002(E)),
Q,1 = characteristic value of leading variable action,
$\gamma_{Q,i}$ = partial factor for accompanying variable action = 1.5,
$\psi_{0,i}$ = 1.0 = factor for category E storage area (see Table A1.1 of BS EN 1990: 2002(E)),
= 0.6 = factor for wind loads (WL) on buildings,
$Q_{k,i}$ = accompanying variable actions.

Table 8.1. Design values of actions for use in accidental and seismic combinations of actions (based on Table A1.3 of BS EN 1990: 2002(E))

Design situation	Permanent condition		Leading accidental or seismic action	A ccompanying variable actions[a]	
	Unfavourable	Favourable		Main	Others
Accidental Equation 6.11a/b	$G_{kj,sup}$	$G_{kj,inf}$	A_d	ψ_{11} or ψ_{21}	$\psi_{2,i} \times Q_{k,i}$
Seismic Equation 6.12a/b	$G_{kj,sup}$	$G_{kj,inf}$	$\gamma_1 \times A_{Ek}$ or A_{Ed}	-	$\psi_{2,i} \times Q_{k,i}$

[a] Variable actions are those considered in Table A1.1 BS EN 1990: 2002(E)

Table 8.2. Design values of actions (STR/GEO) (Set C) (based on Table A1.2(C) of BS EN 1990: 2002(E))

Persistent and transient design situations	Permanent condition		Leading variable action	Accompanying variable actions[a]	
	Unfavourable	Favourable		Main	Others
Equation 6.10	$\gamma_{Gj,sup} \times G_{kj,sup}$	$\gamma_{Gj,inf} \times G_{kj,inf}$	$\gamma_{Q,1} \times Q_{k,1}$	-	$\gamma_{Q,i} \times \psi_{0,i} \times Q_{k,i}$

[a] Variable actions are those considered in Table A1.1 BS EN 1990: 2002(E)

8.4.1 Case 1: For seismic combinations of actions (at the point of splay of the roof beam)

$$\sum\gamma_{Gj} \times G_{k,j} \text{ “+” } A_{Ed} \text{ “+” } \sum\psi_{2,i} \times Q_{k,i} \quad (6.12b)$$

where

$G_{k,j}$ = characteristic moment due to dead loads (on the roof and conveyor) = (15.5 + 15.5) = 31 kN m (previously calculated),
γ_{Gj} = partial factor for dead load = 1.0 for seismic load combination,
A_{Ek} = characteristic moment due to seismic load = (28.5 + 156.2) = 184.7 kN m (see Figs. 8.5 and 8.6),
γ_1 = 1.0,
A_{Ed} = design value of seismic action = $A_{Ek} \times \gamma_1$,
$Q_{k,i}$ = characteristic moments of all variable actions due to LL (on conveyor only) = 36 kN m (see Fig. 8.8),
$\psi_{2,i}$ = 0.0 for roof = 0.8 for storage area,
$Q_{k,i}$ = characteristic moments of all variable actions due to WL (on building) = (326.6 + 57.7) = 383.3 kN m
$\psi_{2,i}$ = 1.0 for WL.

The ultimate design moment for combinations of actions for seismic design situations is:

$$M_{Ed} = \sum G_{k,j} + A_{Ed} + \sum\psi_{2,i}\, Q_{k,I}$$

= + 31 (DL roof and conveyor) + 184.7 (seismic) + $\sum$0.0 × 30.3 (LL roof) + 0.8 × 36.0 (LL conveyor) + 1.0 × 384.3 (WL) = 629 kN m.

The ultimate design thrust for combinations of seismic forces is:

N_{Ed} = (11.03 + 3.9) (DL roof and conveyor) + (7 + 18.9) (seismic) + (0.8 × 9.0) (LL of conveyor) + (38.7 + 7.2) (WL) = 94 kN.

The ultimate design shear for combinations of seismic forces is:

V_{Ed} = (62.8 + 14.7) (DL) + (0.8 × 34.15) (LL) + (4.6 + 25.6) (seismic) + (48.9 + 9.8) (WL) =194 kN.

We considered that during the earthquake the live load on the conveyor and the wind load are likely to exist, with a partial factor $\gamma_{Qk} = 1$ in load combinations.

8.4.2 Case 2: When the seismic force is not acting

The load combinations are:

$\Sigma\gamma_{Gj} \times G_{k,j}$ (DL) + $\gamma_{Q,1} \times Q_{k,1}$ (leading variable LL) + $\gamma_{Q,i} \times \psi_{0,i} \times Q_{k,i}$ (accompanying variable WL)

(see Table A1.2(C), "Design values of actions (STR/GEO) (Set C)", of BS EN 1990: 2002(E)), where

$\gamma_{G,j} = 1.35$; $\gamma_{Q,1} = 1.5$ (LL); $\gamma_{Q,1} \times \psi_{0,i} = 1.5 \times 0.6 = 0.9$ for WL.

The ultimate design moment is:

M_{Ed} = 1.35 × 31 (DL) + 1.5 × 66.3 (LL) + 0.9 × 384.3 (WL) = 423 kN m.

The ultimate design thrust is:

N_{Ed} = [1.35 × (11.03 + 3.9) (DL)] + [1.5 × (22.06 + 9.0) (LL)] + [0.9 × (38.7 + 7.2) (WL)] = 108 kN.

Ultimate design shear

V_{Ed} = [1.35 × (62.8 + 14.7) (DL)] + [1.5 × (126 + 34.15) (LL)] + [0.9 × (48.9 + 9.8) (WL)] = 398 kN.

Therefore a case 1 seismic condition will determine the section of the roof beam.

8.4.3 Design of load combinations for columns

Case 1: When a seismic load is acting but without LL on the roof.
The ultimate design moment in the column at the end of splay is:

M_{Ed} = 1.0 × (104.8 + 36.9) (DL) + 1.0 × (36.8 + 198.0) (seismic) + 1.0 × (336.4 + 61) (WL) + (0.8 × 85.8) (LL conveyor) = 843 kN m.

The ultimate design shear in the column at the end of splay is:

V_u = 1.0 × (11.03 + 3.9) (DL) + 1.0 × (7 + 18.9) (seismic) + 1.0 × (38.7 + 7.2) (WL) + 0.8 × 9.0 (LL) = 94 kN.

The ultimate design thrust in the column at the end of splay is:

$N_u = 1.0 \times (62.8 + 14.7)$ (DL) $+ 1.0 \times (4.6 + 25.6)$ (seismic) $+ 1.0 \times (48.9 + 9.8)$ (WL) $+ (0.8 \times 34.15)$ (LL) $= 193.7$ kN.

Case 2: When a seismic force is not acting.

The ultimate design moment in the column at the end of splay is:

$M_{Ed} = 1.35 \times \Sigma DL + 1.5 \times \Sigma LL + 0.9\Sigma LL + 0.9\Sigma WL = 1.35\,(104.8 + 36.9) + 1.5 \times 209.5 + 0.9 \times 85.8 + 0.9 \times (331.4 + 61) = 935$ kN m

The ultimate design shear at the end of splay is:

$V_{Ed} = 1.35 \times (11.03 + 3.9)$ (DL) $+ (1.5 \times 22.06 + 0.9 \times 9)$ (LL) $+ 0.9 \times (38.7 + 7.2)$ (WL) $= 102.7$ kN.

The ultimate design thrust at the end of splay is:

$N_{Ed} = 1.35 \times (62.8 + 14.7)$ (DL) $+ (1.5 \times 126 + 0.9 \times 34.15)$ (LL) $+ 0.9 \times (48.9 + 9.8)$ (WL) $= 337$ kN.

A case 1 (seismic force combinations) shall be adopted in the design of section of the column (though the moment is slightly less than in case 1).

8.5 Design of section of the column, based on EC 3 and EC 8

Maximum design values at the end of splay in case 1: $M_{Ed} = 843$ kN m
(seismic combinations) $V_{Ed} = 94$ kN
$N_{Ed} = 193.7$ kN

The section will also be checked for case 2: $M_{Ed} = 935$ kN m
(non-seismic combinations) $V_{Ed} = 102.7$ kN
$N_{Ed} = 337$ kN

8.5.1 Initial sizing of the section

Try a section UB 610 × 305 × 179 kg/m; steel grade S 275; $f_y = 275$ N/mm².

Properties of section

Depth of section h = 620.2 mm.
Depth between fillets h_w = 540 mm.
Width of section b = 307.1 mm.
Thickness of web t_w = 14.1 mm.
Thickness of flange t_f = 23.6 mm.
Root radius r = 16.1 mm.
Radius of gyration y–y axis i_y = 25.9 cm.
Radius of gyration z–z axis i_z = 7.07 cm.
Elastic modulus y–y axis W_y = 4940 cm³.
Elastic modulus z–z axis W_z = 743 cm³.
Plastic modulus y–y axis $W_{pl,y}$ = 5550 cm³.
Plastic modulus z–z axis $W_{pl,z}$ = 1140 cm³.
Area of section A = 228 cm².

Note: In accordance with EC 3, the following conventions are followed for a structural member:

$x-x$ axis denotes the axis along a member
$y-y$ axis denotes the major axis of a cross-section of a member
$z-z$ axis denotes the minor axis of a cross-section of a member

8.5.2 Materials

Referring to Clause 6.2 of EC 8:

(1)P The structural steel shall conform to standards referred to in BS EN 1993.
(3a) The actual maximum yield strength $f_{y,max}$ of the steel of *dissipative zones* satisfies the following expression:

$$f_{y,max} \leq 1.1 \times \gamma_{ov} \times f_y$$

where

γ_{ov} = overstrength factor used in design (recommended value = 1.25),
f_y = nominal yield strength specified for the steel grade.

Dissipative zones are expected to yield before other zones leave the elastic range during an earthquake. Thus, for steel grade S 275 (adopted) and $f_y = 275$ N/mm^2,

$$f_{y,max} \leq 1.1 \times 1.25 \times 275 \leq 378 \text{ N/mm}^2.$$

8.5.3 Section classification

Flange

Stress factor $\varepsilon = (235/f_y)^{0.5} = (235/275)^{0.5} = 0.92$.
Outstand of flange $c = (b - t_w - 2r)/2 = (307.1 - 14.1 - 2 \times 16.1)/2 = 130.4$ mm.
Ratio $c/t_f = 130.4/23.6 = 5.52$.
$9 \times \varepsilon = 9 \times 0.92 = 8.28$.

For class 1 section classification, we must have:

limiting value of $c/t_f\ (5.52) \leq 9 \times \varepsilon\ (8.28)$.

So, the condition for class 1 section classification is satisfied.

Web

Ratio $h_w/t_w = 540/14.1 = 38.29$.

Referring to Table 5.2 (sheet 1 of 3) in the Appendix, as the web is subjected to bending and compression, and assuming $\alpha \geq 0.5$ (i.e. more than half of the web is in compression),

$$c/t \leq [(396 \times \varepsilon)/(13 \times \alpha - 1)];\ c/t \leq [(396 \times 0.92)/(13 \times 0.5 - 1)];\ c/t \leq 66.$$

For class 1 section classification, we must have

limiting value of $h_w/t_w\ (38.29) \leq 66$.

So, the condition for class 1 section classification is satisfied.

8.5.4 Moment capacity

Maximum ultimate design moment y–y axis M_{Ed} = 843 kN m
Maximum ultimate shear V_{Ed} = 94 kN
Maximum ultimate thrust N_{Ed} = 193.7 kN

In accordance with Clause 6.2.10 of EC 3, where bending, shear and axial force act simultaneously on a structural member, the moment capacity shall be calculated in the following way:

- Where shear and axial force are present, allowance should be made for the effect of both shear force and axial force on the resistance moment.
- Provided that the design value of the shear force V_{Ed} does not exceed 50% of the design plastic shear resistance $V_{pl,Rd}$, no reduction of the resistance defined for the bending and axial forces in Clause 6.2.9 need be made, except where shear buckling reduces the section resistance.
- Where V_{Ed} exceeds 50% of $V_{pl,Rd}$, the design resistance of the cross-section to a combination of moment and axial force should be calculated using a reduced yield strength $(1 - \rho) \times f_y$ of the shear area, where $\rho = (2V_{Ed}/V_{pl,Rd} - 1)^2$.

(a) *When the web is not susceptible to buckling.*
When the web depth to thickness ratio $h_w/t_w \leq 72\varepsilon$ for class 1 section classification, it should be assumed that the web is not susceptible to buckling, and the moment capacity shall be calculated by the expression $M_{y,rd} = f_y \times W_y$, provided the shear force $V_{Ed} < V_{pl,Rd}$. In our case, we have assumed a section in which $h_w/t_w < 72\varepsilon$. So, the web is not susceptible to buckling.

(b) *When the ultimate shear force $V_{Ed} \leq 0.5 \times V_{pl,Rd}$.*
The ultimate shear at the end of splay V_{Ed} = 100.1 kN and the design plastic shear capacity of the same section is

$$V_{pl,Rd} = A_v \times (f_y\sqrt{3})/\gamma_{Mo}$$

where

γ_{Mo} = partial safety factor on resistance to cross-section = 1,
$A_v = A - 2 \times b \times t_f + 2(t_w + 2r) \times t_f$
$= 22800 - 2 \times 307.1 \times 23.6 + 2 \times (14.1 + 2 \times 16.5) \times 23.6 = 10528\ \text{mm}^2$.

But A_v should not be less than $h_w \times t_w = 540 \times 14.1 = 7614\ \text{mm}^2$, which is satisfied.
Therefore, in our case, $V_{pl,Rd} = [10528 \times (275/\sqrt{3})/1]/10^3 = 1672$ kN, and $0.5 \times V_{pl,Rd} = 0.5 \times 1672 = 836$ kN. So,

$$V_{Ed}\ (94\ \text{kN or}\ 102.7\ \text{kN}) < 0.5 \times V_{pl,Rd}\ (836\ \text{kN}).$$

Therefore, the effect of shear force on the reduction of the plastic resistance moment need not be considered.

(c) *When the ultimate axial force $N_u \leq 0.25 \times N_{pl,Rd}$.*
But, $0.25 \times N_{pl,Rd} = A \times f_y/\gamma_{Mo} = 0.25 \times 228 \times 10^2 \times 275/10^3 = 1568$ kN. In our case, N_{Ed} (193.7 or 337) < 1568 kN. So, the effect on the plastic resistance moment need not be taken into account and the plastic moment capacity is

$$M_{pl,Rd} = f_y \times W_{pl,y} = 275 \times 5550 \times 10^3/10^6$$
$$= 1526\ \text{kN m} > M_{Ed}\ (843\ \text{kN m or}\ 935\ \text{kN m}).$$

Satisfactory

Now, referring to the following expression in EC 8:

$$M_{Ed} = M_{Ed,G} + 1.1 \times \gamma_{ov} \times \Omega \times M_{Ed,E} \qquad (6.6)$$

where

$M_{Ed,G}$ = bending moment in the column due to the non-seismic actions included in the combinations of actions for the seismic design
= [1 × (104.8 + 36.9)] (DL) + [1.0 × (336.4 + 61)] (WL) + [0.8 × 85.8] (LL) = 607.8 kN m,

$M_{Ed,E}$ = design value of bending moment in the column due to design seismic action = [1.0 × (36.8 + 179.0)] (seismic) = 215.8 kN m,

γ_{ov} = overstrength factor = 1.25,

Ω = $M_{pl,Rd,i}/M_{Ed,i}$ of all beams in which dissipative zones are located,

$M_{Ed,i}$ = design value of bending moment in beam i in the seismic design
= 629 kN m,

$M_{pl,Rd,i}$ = $W_{pl,y} \times f_y/\gamma_{Mo}$ = 5550 × 10^3 × 275/106 = 1526 kN m,

Ω = 1526/629 = 2.43.

Therefore

$$M_{Ed} = 607.8 + 1.1 \times 1.25 \times 2.43 \times 215.8 = 1329 \text{ kN m}.$$

and so

$M_{pl,Rd}$ (1526 kN m) > M_{Ed} (1329 kN m). Satisfactory

8.5.5 Shear buckling resistance

Shear buckling resistance need not be checked if the ratio $h_w/t_w \le [396\varepsilon/(13\alpha - 1)]$. In our case,

$$h_w/t_w = 540/14.1 = 38.29 \text{ and } 396\varepsilon/(13\alpha - 1) = 66 \text{ (already calculated)}.$$

So,

$$h_w/t_w\ (38.29) < 396\varepsilon/(13\alpha - 1)\ (66).$$

Therefore, shear buckling resistance need not be checked.

8.5.6 Buckling resistance to compression

The buckling resistance to compression is given by

$$N_{b,Rd} = \chi \times A \times f_y/\gamma_{M1}$$

where

χ = reduction factor = $1/(\Phi + (\Phi^2 - \bar{\lambda}^2)^{0.5})$,

Φ = $0.5[1 + \alpha \times (\bar{\lambda} - 0.2) + \bar{\lambda}^2]$,

$\bar{\lambda}$ = $L_{cr}/(i_y \times \lambda_1)$,

L_{cr} = buckling length in the major axis = 0.85 × L = 0.85 × 10.5 = 8.93 m,
(*Note:* 0.85 factor is taken because the bottom is assumed hinged and the top is assumed fixed),

λ_1 = 93.9ε = 93.9 × 0.92 = 86.4,

i_y = 25.9.

Therefore,

$\overline{\lambda}$ = 8.93 × 10³/(25.9 × 10 × 86.4) = 0.4;
h/b = 620.2/307.1 = 2 > 1.2.

Referring to Table 6.2 ("Selection of buckling curve for a cross-section") of EC 3, and following the buckling curve "a" of Fig. 6.4 of EC 3 (see Appendix), with $\overline{\lambda} = 0.4$ and $\chi = 0.95$, the buckling resistance to compression is therefore

$N_{b,Rd}$ = 0.95 × 228 × 10² × 275/10³
= 5957 kN >> N_{Ed} (193.7 kN). Satisfactory

Now, referring to the following expression given in EC 8:

$$N_{Ed} = N_{Ed,G} + 1.1 \times \gamma_{ov} \times \Omega \times N_{Ed,E} \tag{6.6}$$

where

$N_{Ed,G}$ = compressive force due to non-seismic action included in the combination of actions for the seismic design situation
= [1.0 × (62.8 + 14.7)] (DL) + [1.0 × (48.9 + 9.8)] (WL) + [0.8 × (34.15)] (LL)
= 164 kN,

$N_{Ed,E}$ = compression force in the column due to the design seismic action
= (4.6 + 25.6) kN = 30.2 kN (see Clause 21.3.2, "Design of load combination for column"),

Ω = the minimum value = $N_{pl,Rd,i}/N_{Ed,i} = (A \times f_y/\gamma_{Mo})/N_{Ed,i}$ of all beams in which dissipative zones are located,

$N_{Ed,i}$ = design value of the compression in beam i in the seismic design situation
= 94.4 kN (see Clause 21.3.1, case 1 for seismic combinations of actions),

$N_{pl,Rd,i}$ = corresponding plastic compression = 228 × 10² × 275/1.0/10³ = 6270 kN,

Ω = 6270/94.0 = 66.7,

γ_{ov} = overstrength factor = 1.25 (recommended, see Clauses 6.1.3(2) and 6.2(3)).

Therefore

N_{Ed} = 164 + 1.1 × 1.25 × 66.7 × 32.2 = 164 + 2953 = 3117 kN

and

$N_{pl,Rd}$ = 6270 kN.

So, we have

$N_{Ed}/N_{pl,Rd}$ = 3117/6270 = 0.497 < 1.0. Satisfactory

Again, referring to the following expression given in EC 8:

$$V_{Ed} = V_{Ed,G} + 1.1 \times \gamma_{ov} \times \Omega \times V_{Ed,E} \tag{6.6}$$

where

$V_{Ed,G}$ = shear force in the column due to the non-seismic actions included in the combinations of actions for the seismic design situation
= (11.03 + 3.9) (DL) + (7 + 18.9) (WL) + (0.8 × 9) (LL) = 48 kN,

$V_{Ed,E}$ = shear force in column due to the design seismic action
= 7.0 + 18.9 = 25.9 kN,

Ω_i = $V_{pl,Rd,i}/V_{Ed,i}$ = plastic shear resistance in beam/design shear in beam
= 1672/194 = 8.6 (plastic shear resistance and design shear in beam were calculated previously).

Therefore

$V_{Ed} = 48 + 1.1 \times 1.25 \times 8.6 \times 25.9 = 354$ kN

and hence

$V_{Ed}/V_{pl,Rd} = 354/1672 = 0.2 < 1.0.$ Satisfactory

8.5.7 Buckling resistance moment

The column is assumed to be hinged at the base and fixed at the top. The outer compression flange in contact with the side rails spaced at 2 m intervals is restrained against buckling. The inner compression flange, however, is not restrained against buckling. So, a horizontal strut member halfway along the column is provided to restrain the compression flange against buckling. Therefore, the unsupported length of the compression flange against buckling = 10.5/2 = 5.25 m.

The buckling resistance moment is calculated in the following way. Referring to Clause 6.3.2.4 of EC 3, "Simplified assessment methods for beams with restraints in buildings", members with lateral restraint to the compression flange are not susceptible to lateral torsional buckling if, given the length L_c between restraints, the resulting slenderness $\bar{\lambda}_f$ of the equivalent compression flange satisfies:

$$\bar{\lambda}_f = k_c \times L_c/(i_{f,z} \times \lambda_1) \leq \bar{\lambda}_{c0} \times (M_{c,Rd}/M_{y,Ed})$$

where

k_c = a slenderness correction factor for moment distribution between restraints = 0.90,
W_y = section modulus against the major axis = 5550 cm^3,
f_y = yield strength of steel of grade S 275 = 275 N/mm^2,
γ_{M1} = partial safety factor to resistance of a member to instability = 1,
$M_{c,Rd} = W_y \times f_y \times 1.1 \times \gamma_{ov}/\gamma_{M1} = 5550 \times 10^3 \times 275 \times 1.1 \times 1.25/10^6 = 2099$ kN m (for $f_{y,max} = 1.1 \times \gamma_{ov} \times f_y = 1.1 \times 1.25 \times f_y = 1.375 \times f_y$ and γ_{ov} = overstrength factor – see Clause 6.2 of EC 8),
$M_{y,Ed}$ = 843 kN m,
L_c = 525 cm,
$i_{f,z}$ = radius of gyration of the equivalent compression flange composed of the compression flange plus 1/3 of the compressed part of the web area, about the minor axis of the section = 8.11 cm,
$\lambda_1 = 93.9 \times 0.92 = 86.4$,
$\bar{\lambda}_{c0}$ = slenderness limit of the equivalent compression flange defined above $= \bar{\lambda}_{LT,0} + 0.1 = 0.4 + 0.1 = 0.5$ (see Clause 6.3.2.3 of EC 3),
$\bar{\lambda}_f = k_c \times L_c/(i_{f,z} \times \lambda_1) = 0.9 \times 525/(8.11 \times 86.4) = 0.67.$

Thus

$\bar{\lambda}_{c0} \times (M_{c,Rd}/M_{y,Ed}) = 0.5 \times 2099/1329 = 0.79 > 0.67,$

so no necessity of reduction of the design buckling resistance moment. Therefore,

$M_{Ed}/M_{b,Rd} + N_{Ed}/N_{b,Rd} = 843/2099 + 193.7/5957 = 0.40 + 0.03 = 0.43 < 1.0$

where

1.1 × 1.25 (overstrength factor of steel) = maximum yield strength factor of steel of dissipative zones (refer to Clause 6.2, "Materials", of EC 8).

Satisfactory

Therefore adopt UB 610 × 305 × 179 kg/m for the column (see Fig. 8.9).

8.6 Design of section of the roof beam

At the end of splay:

M_{Ed} = 596 kN m,
V_{Ed} = 192 kN,
N_{u} = 92.4 kN.

As the ultimate moment, shear and thrust are less than those in the column, adopt the same section of the roof beam, i.e. UB 610 × 305 × 179 kg/m for the roof beam. See Fig. 8.9.

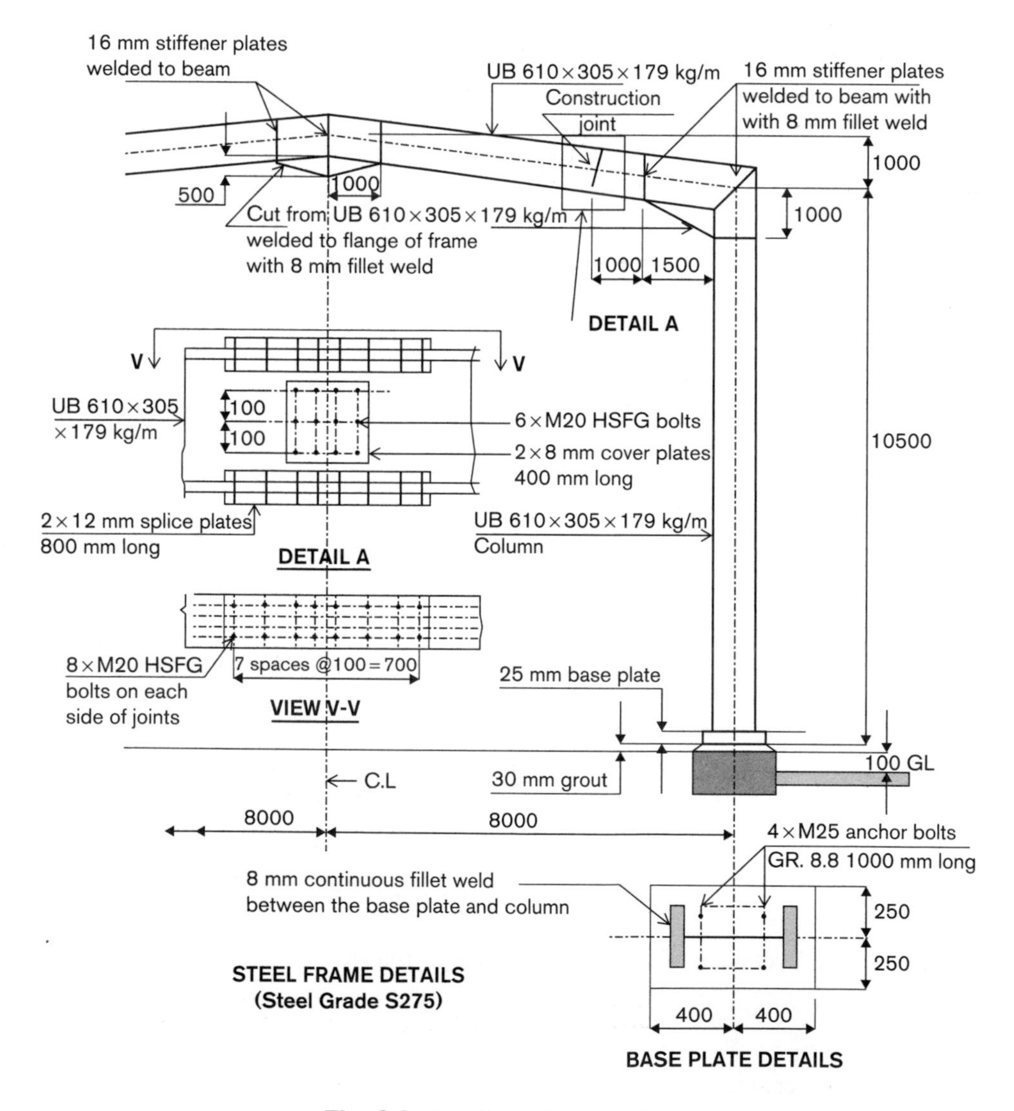

Fig. 8.9. Details of the portal frame

References

Eurocode, 2002a. BS EN 1991-1-1: 2002(E), Eurocode 1. Actions on structures. General actions. Densities, self-weight, imposed loads for buildings.

Eurocode, 2002b. BS EN 1990: 2002(E), Basis of Structural Design.

Eurocode, 2004. BS EN 1998-1: 2004, Eurocode 8. Design of Structures for Earthquake Resistance.

Eurocode, 2005. BS EN 1993-1-1: 2005, Eurocode 3, Design of Steel Structures.

Eurocode, 2006. BS EN 1991-1-4: 2006, Eurocode 1, Actions on Structures. General actions. Wind.

Kleinlogel, A., 1958. *Rahmenformeln "Rigid Frame Formulae"*. Wilhelm Ernst & Sohn, Berlin.

PART II

ANALYSIS AND DESIGN OF A STEEL-FRAMED STRUCTURE SUBJECTED TO MACHINE VIBRATIONS

This part of the book addresses the design of structures to resist dynamic loadings – either from supported machinery or from any other external sources. A structure must satisfy the criteria for static loadings and for resisting the dynamic loadings induced by vibrations due to unbalanced rotating mass. This part of the book is concerned mainly with the dynamic effects of the machine – namely a turbo-generator – on the supporting structure and also on the foundation of the supporting structure.

CHAPTER 9

General Principles and Practices

9.1 Brief description of the structure

This chapter addresses a case study of the dynamic effects of rotating parts of a turbo-generator machine (during operation) mounted on an elevated structural steel frame (see Figs. 9.1 and 9.2). The structure, supporting a 10 MW capacity turbo-generator unit in a power plant, is 11.0 m long, 5.0 m wide and 5 m high above ground level. The turbo-generator unit is installed on 1100 mm deep steel grid beams at 5 m above ground, with overhang (turbine floor) for the maintenance of the unit.

The structure is founded on a 2.5 m thick raft at a depth of 3.75 m below ground floor level. A clear separation joint of 40 mm has been kept all round the structure at ground floor level in order to avoid the transmission of mechanical vibrations arising from the

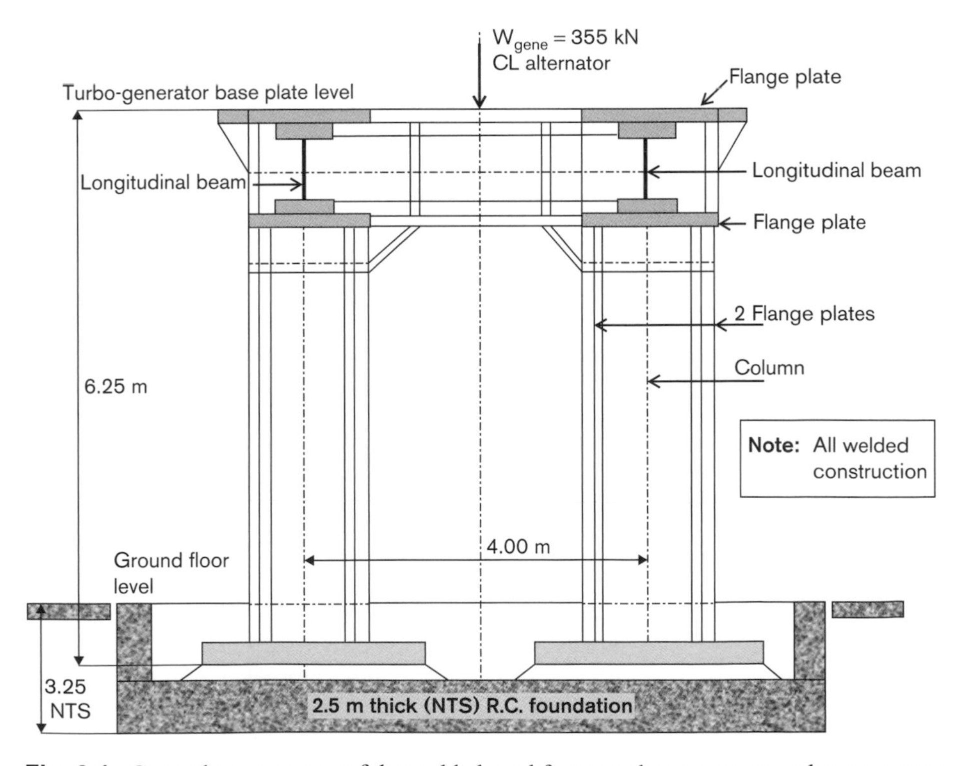

Fig. 9.1. General arrangement of the welded steel framework to support a turbo-generator

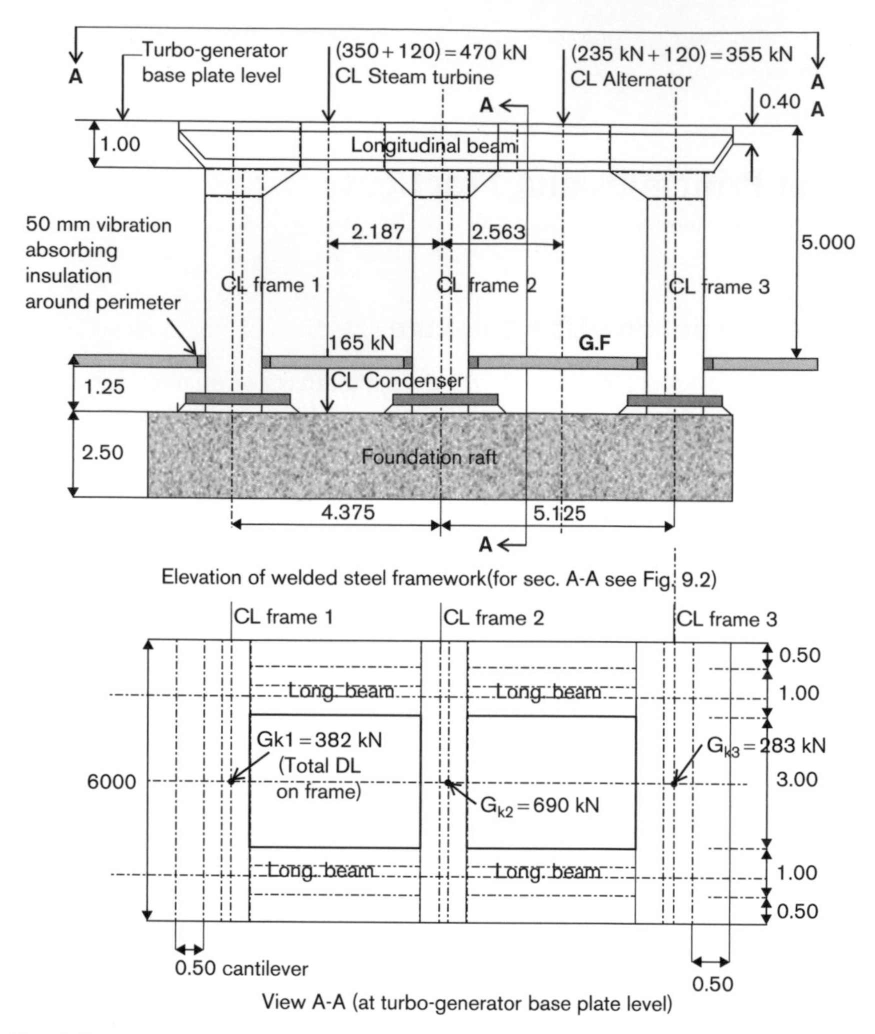

Fig. 9.2. Section A-A: Turbo-generator support frame 2 (for plan view see Fig. 9.1)

turbo-generator unit. The gap of 40 mm has been filled with vibration-absorbing material (see Figs. 9.1 and 9.2).

9.2 Functional aspect

The structure supports the turbo-generator unit and is designed to resist the dynamic vertical and horizontal forces due to the induced vibration generated during the rotational movement of machines parts. The turbine floor around the machine is provided with adequate space for everyday observation and maintenance of the unit.

9.3 Choice of open or covered top

It is not mandatory to cover the units. In many industrial projects the turbo-generator units are installed on open ground without any cover. In our case, the turbo-generator is housed in the powerhouse building to protect against bad weather conditions and to provide better facilities for the operational personnel (see Figs. 9.3 and 9.4).

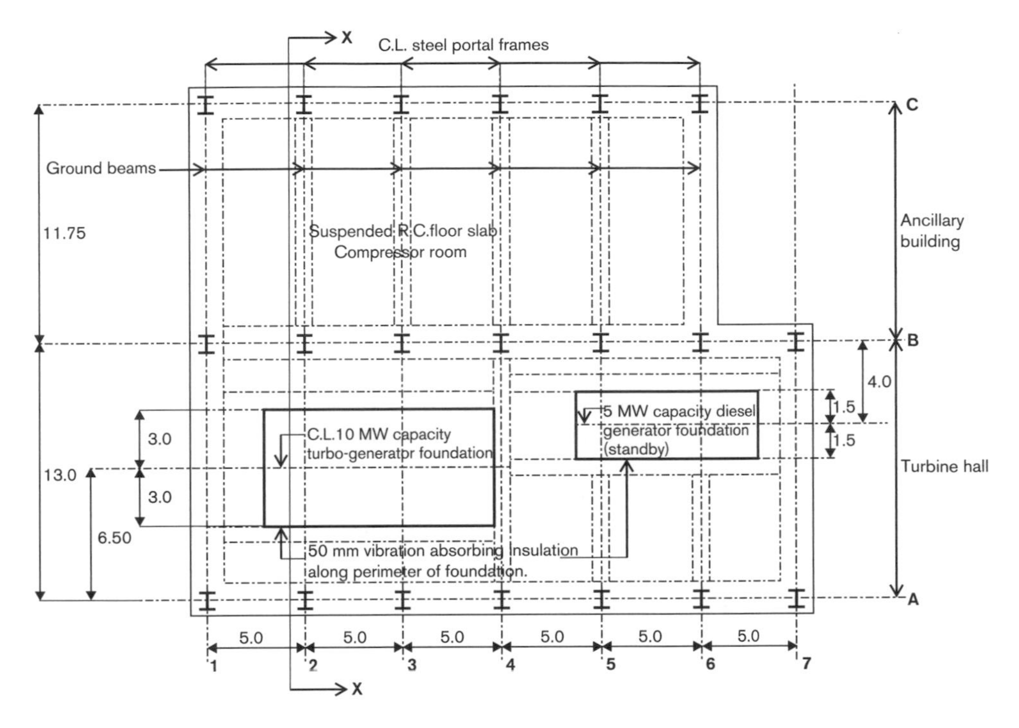

Fig. 9.3. Ground floor plan of the turbo-generator building

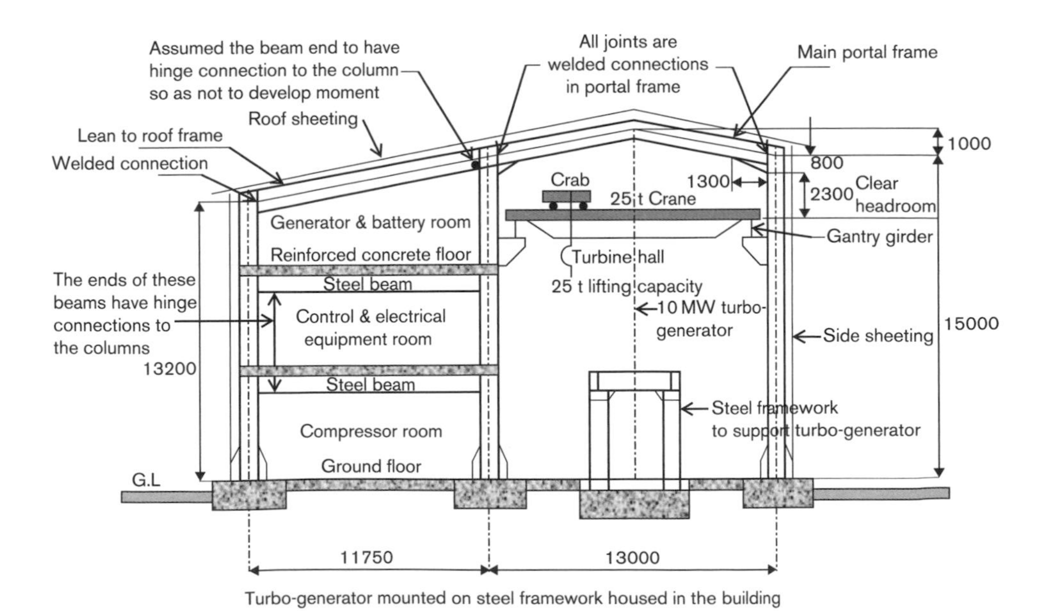

Fig. 9.4. Section X-X of the turbo-generator building

9.4 Selection of construction material

The turbo-generator set is mounted mainly on reinforced concrete or on a steel portal framing system. The transverse frames carry the vertical and horizontal dynamic loads generated from the rotating mass (rotors) of the unit, in addition to dead loads of the machine and self-weight of the structure. The transverse frames are connected rigidly to the edge longitudinal beams, which act as longitudinal frames to transfer any unbalanced excess loads in the transverse frames.

A reinforced concrete framing system with mass taken generally equal to 10 to 15 times the mass of the rotors is highly overtuned so that the vertical natural frequency (cycles per minute (c.p.m.)) of the frames is significantly higher than the operating speed of the machine. Thus, there is a rare possibility of resonance of the structure occurring.

A steel portal framing system is also used to support the turbo-generator machine. The steel framing system has the same arrangement of structural members. There are three main transverse portal frames (or more as the capacity of machine increases), which support the turbo-generator unit. The advantage of a steel-framed supporting structure is that the frames can be individually fabricated at the factory, without depending on the weather conditions, and then be assembled at the site in the minimum amount of time.

The natural frequencies (even the first harmonics) are lower, both in the vertical and in the horizontal plane, than the operating speed of the machine, which means that the entire structure is undertuned and there is a possibility of resonance occurring at the start-up and stopping of the machine (with a normal operating speed range between 1500 and 3000 r.p.m.). This resonance may cause the magnification of vibrations of the structure, which may result in damage to the structure due to the stressing of the structural elements beyond permissible limits, and the structural material may reach a fatigue state. As a prerequisite, to avoid resonance occurring the natural frequency of the supporting structure of the turbo-generator should differ from the operating speed of the machine by at least ±20%; at the same time, the amplitude of vibration should be limited to a permissible value of 0.02–0.03 mm for the turbine operating at 3000 r.p.m. and above. Moreover, the steel framing system forms the soft bedding of the machine base and this ensures the rotation of the rotors about the centre of gravity, thus reducing the unbalanced dynamic force and guaranteeing longer life for the supporting bearings than for a reinforced concrete support. Because of the soft bedding, the vibrations transmitted to the soil and to the adjacent foundations are reduced.

Any unforeseen vibrations during operation can be dampened out by stiffening the framework, thus adopting a rigid frame portal construction, and also by the installation of shock absorbers under the base of machine, if the situation arises.

The estimated quantity of steel required for 10 MW is between 50 and 60 tonnes, which is significantly less than if reinforced concrete were used.

Considering the above points, we conclude that the steel-framed construction is the best option to adopt in this case (see Figs. 9.1 and 9.2).

9.5 Loads on the turbo-generator supporting structure

Dead loads

- Self-weight of the supporting structure.
- Weight of the machine, including weight of the rotors.
- Weight of condenser, oil, air cooler, tank, piping, etc.

Table 9.1. Approximate weight of rotors of turbo-generators at various outputs (based on Table 33 by Major, 1962)

Output (MW)	Total weight of turbo-generator including rotors (kN)	Total weight of rotors of turbine and generator (kN)	Weight of rotors (% of the total weight)
5	620	120	19.5
10	1000	186	18.6
25	1800	315	17.5
50	3000	500	16.6
100	5000	830	16.0

Live loads

- Vertical dynamic loads due to faulty balancing of mechanical equipment.
- Horizontal dynamic loads acting normal to the longitudinal axis and due to the unsatisfactory balancing of the machine, caused by the centrifugal force generated from the revolving eccentric masses.
- The longitudinal horizontal dynamic loads.

Note: For the computation of dynamic effects, the weight of rotors and their points of application are of greatest importance.

Notes: (1) Normally, the manufacturer provides the design data.
(2) Nowadays much higher capacities (up to 500 MW, or even 1000 MW) of machines are installed.
(3) The weight of rotating parts is always transmitted through the bearings, not through the points of support of the stationary parts. The first bearing of the turbine is supported on the first left structural frame, the second bearing of the turbine and the first bearing of the generator are supported on the second structural frame from the left. The second bearing of the generator is supported on the third structural portal frame from the left.
(4) The weight of the generator rotor amounts to 60% of the total weight of all rotating masses.

In our case, the manufacturer has supplied the machine loads for a 10 MW capacity turbo-generator as follows:

Weight of steam turbine = 350 kN
Weight of reduction gear = 50 kN
Weight of generator = 235 kN
Weight of bed frame = 150 kN
Weight of condenser = 142 kN
Weight of auxiliaries = 40 kN

Total = 967 kN

We assume that the total weight acting in line with the turbine weight at the top frame level is

W_{tur} = 350 (turbine) + 50/2 (reduction gear) + 150/2 (bed frame) + 40/2 (auxiliaries) = 470 kN and the total weight acting in line with the alternator (generator) weight at the top frame level is

$$W_{gene} = 235 \text{ (alternator)} + 50/2 \text{ (reduction gear)} + 150/2 \text{ (bed frame)} + 40/2 \text{ (auxiliaries)} = 355 \text{ kN.}$$

The *dynamic effect* is a function of time. Some dynamic forces act regularly according to certain constant laws of sine and cosine and follow the changes in the speed of the machine. Moreover, there are impulsive, shock-like forces which act irregularly. These forces due to the operation of generator are the dynamic loads. They occur as a result of sudden electrical shocks in the generator and may be many times greater than the normal forces.

Due to the mutual magnetic effect between the stator and the rotor, a shock in the form of a couple may be generated and tend to break the stator off the frame support. The moment arising from this couple is called a *short-circuit moment*. Owing to the short-circuit moment, a vertical load P acts on the transverse beam at the points of support of the stator. The magnitude of this load is

$$P = M/l$$

where

M = short-circuit moment,
l = distance of the supporting points of the stator.

The load P acts vertically upwards and downwards and is generally specified by the manufacturer. In the absence of manufacturer data, the short-circuit moment may be estimated by the following expression,

$$M = 4 \times W$$

where W = capacity of the generator in MW.

References

Major, A., 1962. *Vibration Analysis and Design of Foundations for Machines and Turbines*, Collet's Holdings, London.

CHAPTER 10

Design Concept

10.1 Method of computations

The steel structure supporting the turbo-generator machine is subjected to dynamic loads causing vibrations during operating conditions. Before proceeding to the details of structural design, it is necessary to analyse the dynamic effects and behaviour on the structure.

The *dynamic loads* may be defined as the forces generated by moving periodic or impact loads acting on the machine mounted on the structural framework. Dynamic loads are usually characterised by their regular recurrence at certain intervals of time, whereas their magnitude changes rapidly with time.

In general, there are two computation methods for studying the effects and behaviour of a structure subjected to dynamic loads:

(1) The resonance method: This method follows the principles of the theory of vibration, considering the natural frequencies of the structural framework that supports the turbo-generator machine. The natural frequencies of the structural elements of the framework are independent of the generating forces of the machine and depend entirely on the mass and shape as well as the mechanical properties of material (steel in our case). According to this method, the most important requirement is that the structure should be out of tune, i.e. all the structural elements of the supporting structural framework should be analysed and designed to ensure that the natural frequency differs from the machine speed by at least 20% to avoid the occurrence of resonance of the supporting framework.

(2) The amplitude method: The fundamental requirement is that the amplitude of vibration of structural elements of the supporting framework should not exceed a certain permissible value. Different permissible amplitudes are prescribed for different types of machinery.

10.1.1 Resonance method

In this section we shall discuss in detail the resonance method based on experiments to determine the generating forces, focusing on the dynamic effect of the centrifugal force.

Resonance problems involving the transient resonance occurring with deep undertuning have been studied in the analysis of resonance phenomena. Practical investigations of vibrations occurring during operating conditions of machines have shown that, where the natural frequency of the supporting structure is lower than the operating speed of the machine (as always in the case of horizontal frequencies at operating speeds of from 1500 to 3000 r.p.m.), maximum forces will occur even during the short period of acceleration and deceleration when the machine is started or stopped. These forces are significant for the design of frame elements subject to bending. Results of synchronised vibration measurements by Major (1962) showed that the peak amplitude of a deeply undertuned element is much higher during acceleration than at the normal operating speed. The corresponding dynamic forces are also much higher.

The analysis and design of a steel framework to support the turbo-generator machine is troublesome because the vertical natural frequency is usually lower than the operating speeds of the machine. The natural frequency of a turbine supporting structure cannot be determined with the necessary precision in practice because of the varied and numerous factors involved, some of which can not be estimated accurately.

The natural frequency is computed based on the deformation of the individual elements of the structural framework: the beams and the columns. When the natural frequency computed on the basis of deformation of the frame elements differs considerably from the natural frequency computed on the basis of the elasticity of bedding alone, then the natural frequency computed by a simultaneous consideration of the deformations of the supporting structure and the elasticity of the bedding material will differ only slightly from the natural frequency computed from the deformation of the soil only.

The natural frequency of an absolutely rigid body computed on the basis of the elasticity of bedding (soil) alone will be between 300 and 800 vibrations per minute (c.p.m.), and differ considerably from the number of revolutions per minute (r.p.m.) of high-speed machines (1500 to 3000 r.p.m.). For this reason the elasticity of the bedding (soil) may be neglected in computations in normal operating conditions. With high-speed machines of about 3000 r.p.m. the natural frequencies calculated based on the deformation of the supporting elements (usually 2000 to 4000 c.p.m.) can be taken into account at normal operating speed.

In making a decision for the inclusion of the elasticity of the bedding in the analysis of natural vibrations, the following practice justified by experiments in Hungary may be recommended:

- With machines operating at normal speeds of 1500 r.p.m. and above, the natural frequency of the supporting structure can be computed without considering the elasticity of bedding material. As a check, the elasticity of the bedding can be taken into account in view of transient resonance – the natural frequency should be determined for this particular case too.

The vibration measurements of a machine supporting structure were carried out by the resonance method by Major (1962) for the synchronised measuring of the vibrations. The vibration diagram (see Fig. 10.1) showed that at 700 r.p.m. the amplitude has a peak value A_1. The natural frequency, calculated based both on the deformation of the supporting structure and on the elasticity of bedding material was also nearly 700 c.p.m. and the peak value is an indication of resonance.

The vibration diagram also showed that at 1400 r.p.m. a second peak amplitude value A_2 (though smaller) was observed, which was closely connected to the critical speed of the turbine rotating shaft during operating conditions. The amplitude measured at the normal operating speed 3000 r.p.m. of the machine was A, which was smaller than A_1 and A_2. The effects of resonance that occurred at 700 r.p.m. having been transmitted by the soil could be observed at several structures in the vicinity of the turbine support.

From experiments carried out independently by different people, we may arrive at the conclusion that for machines operating at 1500 r.p.m. or higher, the natural frequency is found to be less than 800 r.pm. when both the elasticity of the framed structure and the elasticity of the soil are taken into account.

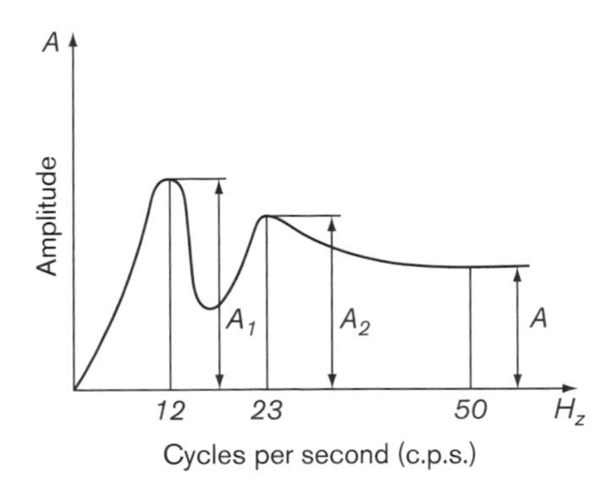

Fig. 10.1. Amplitude curve from synchronized vibration measurements (after A. Major, 1962)

10.2 Stresses due to maximum dynamic effect

The framed structure for a turbo-generator support should be designed for mechanical strength and safety against vibrations. From the strength point of view, the stresses due to maximum dynamic effect should be lower than the permissible stresses.

For this reason, when the natural frequency of the framed structure (N_e in c.p.m.) is lower than the operating speed of machine (in r.p.m.), the maximum dynamic forces at the starting or the stopping condition (defined as *transient state*) occur on the structure and are highly significant for the structural strength concerned. This condition can be justified due to the fact that some machines are very frequently started and stopped (sometimes many times per day, if not per hour).

The turbines, on the other hand, are started very gradually in order to attain the required "starting time" necessary to warm up the machine. This process usually takes 20 to 40 minutes. The "stopping time" may even be longer than the starting time (40 to 60 minutes). During this process we should remember that the rotor of the machine will rotate within ±5% of all frequencies for 1 to 2 minutes. With deep undertuning of a structure, a resonance may therefore occur for a few minutes, as the natural frequency of the structure may coincide with the speed of the machine. Hence the dynamic forces generated due to resonance causing vibration may overstress the structure by exceeding permissible stresses, and have a serious damaging effect on the structure, even though they are transient. So, the dynamic forces should not be neglected, in spite of their transient nature.

Thus, for $N_e < N_o$ during the starting and stopping periods, the conditions are significant for the design of the structure, as well as for $N_e > N_o$ during normal operating conditions.

10.3 Amplitudes

As far as vibration is concerned, the amplitudes at normal operating speeds are of highest importance, as opposed to those at starting or stopping. This is because the amplitudes that may occur at starting or stopping exist for a very short period of time and at speeds lower than the normal operating speed. For this reason they do not seem to be dangerous. Moreover, at starting, the machine can be accelerated quickly over the resonance range. Deceleration, however, may cause larger amplitudes to remain for extended periods and in some cases cannot be neglected.

The amplitudes occurring at normal operational speeds must be considered significant because at this stage the machine is usually working under full loads. Accordingly, the amplitudes at normal operating speeds should be lower than the permissible value, irrespective of whether the natural frequency of the framed structure is smaller or higher than the speed of the machine.

The simplest vibrating system, consisting of a single mass, is usually assumed for the analysis. Actually, turbine-generator machines are multi-mass systems and this condition should be therefore considered.

The amplitudes of a single-mass vibrating system follow the curve as shown in Fig. 10.2, whereas in a multi-mass vibrating system at normal operating speed the amplitude may exceed considerably the operating amplitude as shown in Fig. 10.3. by considering the single-mass system it is easy to visualise the structure but not accurate due to multi-mass vibrating system.

Despite its limitations, a framed structure for a turbo-generator machine can be designed satisfactorily, based on many experiments. The results of numerous tests carried out to determine the maximum operating amplitude have shown that the resonance curve varies rather "capriciously" after the first peak and does not resemble in any respect the resonance curve of one- or two-mass vibrating systems. The first peak is generally followed by a continuous reduction or, at most, by a very slight increase of the amplitudes. Thus, the maximum amplitudes do not deviate considerably from those occurring during acceleration or deceleration. Therefore, the amplitudes at acceleration and deceleration respectively should be considered significant in the analysis of the framed structure of a turbo-generator machine, assuming that A_{20} is nearly equal to $A_{1\max}$ as shown in Figs. 10.2 and 10.3.

Figs. 10.2 and 10.3 should be considered when the natural frequencies are substantially lower than the operating speed of the machine, i.e. in the case of deep undertuning.

When determining the permissible amplitudes the vibrations occurring should be investigated first. Results of relevant experiments have produced the permissible amplitudes shown in Table 10.1. The lower value of the range applies to machines operating at the higher speed of 3000 r.p.m.

The results of further experiments on machines running at different speeds produced the permissible values of amplitudes shown in Table 10.2. Experimental values closely compared to the computed values.

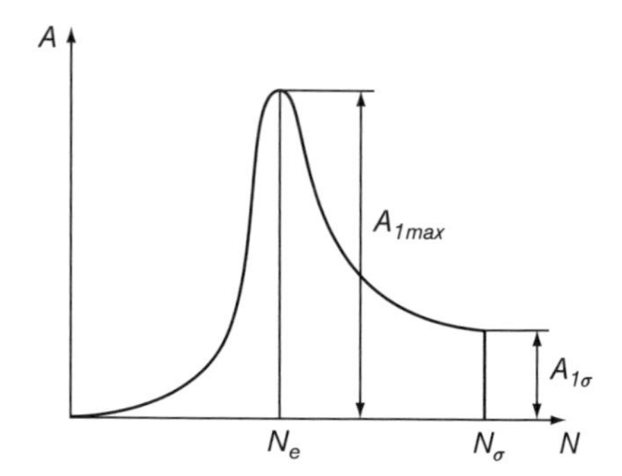

Fig. 10.2. Amplitudes of a single-mass system plotted against machine speed (after Major, 1962)

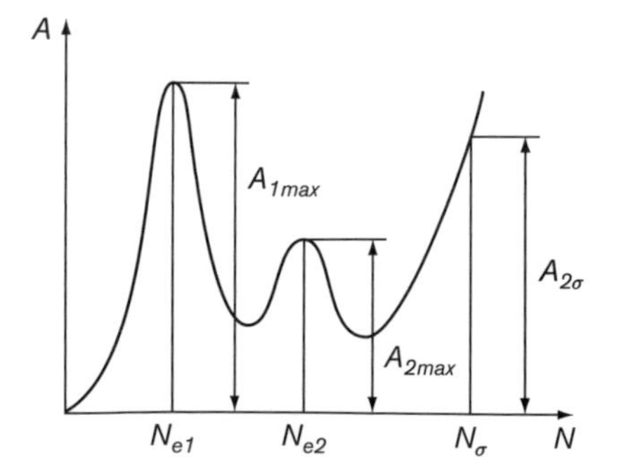

Fig. 10.3. Amplitudes of a multi-mass system plotted against machine speed (after Major, 1962)

Table 10.1. Balancing and operating amplitudes (based on figures by Major, 1962)

1. With excellent balancing	A = 0.02–0.04 mm
2. With good balancing	A = 0.04–0.05 mm
3. With fair balancing	A = 0.06–0.10 mm
4. With unsatisfactory balancing	A = 0.10–0.16 mm

Table 10.2. Permissible operating amplitudes for different speeds (based on figures by Major, 1962)

Machine r.p.m.	Permissible range of amplitude
3000 or above	0.02–0.03 mm
1500	0.04–0.06 mm
750 or below	0.08–0.12 mm

10.4 Dynamic effect of the centrifugal force

An unsatisfactory balancing of the rotating parts of a machine gives rise to vibrations due to the dynamic effect of the centrifugal force. This may be further explained by the fact that when the centre of gravity of the rotating mass of a rotor deviates from the centre of the geometrical axis of rotation of the shaft due to unsatisfactory balancing, then centrifugal forces acting at the axis originate during rotation of the rotor.

Since the rotation is fast and the weight of the mass is considerable, even the smallest eccentricity between the centre of gravity and the axis of rotation generates considerable centrifugal force and the dynamic effect of the centrifugal force results in the creation of a dynamic force on the structural framework. For example, for a rotor of 50 kN weight, an eccentricity of 0.2 mm will generate a centrifugal force of 100 kN at a machine speed of 3000 r.p.m.

These forces are mainly the ones occurring in a turbine supporting framework due to the unsatisfactory balancing of rotating parts, and being centrifugal forces, they produce dynamic effects. The centrifugal force C may be expressed by the following equation (based on equation 588 by Major, 1962):

$$C = m \times e_1 \times \omega^2 \qquad (1)$$

where

m = rotating mass = G_r/g,
G_r = weight of rotor,
g = gravitational acceleration,
e_1 = radius of rotation of the centre of gravity of the rotating mass, i.e. eccentricity,
ω = angular velocity in radians/sec.

The angular velocity can be expressed in terms of speed N. So,

$$\omega = 2 \times \pi \times N/60.$$

Inserting this value in equation 1, we obtain

$$C = G_r \times e_1 \times N^2 \times (2\pi/60)^2/g = B \times G_r \times e_1 \times N^2 \qquad (2)$$

where

N = number of revolutions per minute,
e_1 = eccentricity in cm,
G_r = weight of rotor in tonnes,
$B = 1.1 \times 10^{-5}$.

Referring to equation 2, the centrifugal force is proportional to the square of speed, and reaches its maximum value C_{max} at the operating speed. The centrifugal force at an intermediate speed N_o can be expressed as $C = C_{max} \times N^2/N_o^2$.

The value of the centrifugal force determined above must be multiplied by a *dynamic factor* to get the dynamic effect. So, we have (based on equation 592 by Major, 1962):

$$Z = C \times \upsilon \qquad (3)$$

where

Z = theoretical dynamic load,
C = static effect of the centrifugal force,
υ = dynamic factor.

The value of the dynamic factor may be expressed by the following equation (based on equation 593 by Major, 1962):

$$\upsilon = 1/[(1 - N^2/N_{e2)})^2 + (\Delta/\pi)^2 \times (N^2/N_{e2)})]^{0.5} \qquad (4)$$

where

N = speed of the machine,
N_e = natural frequency of the structure,
Δ = logarithmic decrement of damping.

The maximum value of the dynamic factor will be obtained by differentiating equation 4 with respect to N

$$N = N \times [1/2 \times (\Delta/\pi)^2)]^{0.5}. \qquad (5)$$

In practice, the maximum value of the dynamic factor occurs when $N = N_e$, i.e.

$$\upsilon_{max} = \pi/\Delta \text{ or } \upsilon_{max} = 2\pi/\psi \qquad (6)$$

(based on equation 595 by Major, 1962), where ψ = specific internal resistance of the elastic connection of a vibration damper.

From the above expressions we find that the dynamic factor is inversely proportional to damping: the higher the damping, the lower the dynamic effect. The quantity ψ is characteristic of different building material and depends on the stress distribution in the material of the structure, as well as the magnitude of the stress. It is mostly affected by the shear stress, but also by the duration of loading and the temperature. Tables 10.3 and 10.4 give values of ψ determined experimentally for various building materials and structures respectively.

10.5 Critical speed

When a rotating elastic mass, such as the prismatic beam or shaft of a machine, suffers a displacement perpendicular to its axis due to any kind of impulse, the elastic reaction returns the beam to its initial position and elastic oscillations are set up. Since, on the one hand, these vibrations are the result of the eccentricity of the beam (shaft) and of the

Table 10.3. Experimental values of ψ for various materials (based on Table 34 by Major, 1962)

Material	ψ
High-tensile steel	0.018
Tool steel	0.015
Machine steel	0.010
Ball-bearing steel	0.010
Magnetic steel	0.023
Pure copper	0.33
Natural cork	0.038
Rubber	0.24
Pine	0.12

Table 10.4. Experimental values of ψ for various structures (based on Table 35 by Major, 1962)

Types of structure	ψ (average)
Steel bridge	0.17
Nailed wooden beam with diagonal web	0.30
Reinforced concrete beams	0.56
Reinforced concrete frame	0.30

rotating mass attached to it, and, on the other hand, they are due to the elastic deflection, they are *forced transverse vibrations*. The equation of this motion can be derived from the equation of harmonic motion.

The *critical speed* may be computed from the simple static deflection (Δ). Thus, consider a shaft simply supported at the ends on bearings and a rotor mass fixed at the centre of the shaft (see Fig. 10.4).

Let,

W = weight of rotating mass acting at the centre of the shaft,
L = span of the shaft between the bearing supports,

then, the deflection at the centre of the shaft is

$$\Delta = WL^3/6EI$$

where

E = elasticity modulus of steel = 21×10^7 kN/m^2,
I = moment of inertia of the section of the shaft,
N = critical speed of the rotor = $30/(\Delta)^{0.5}$ r.p.m.

Since the rotating parts of the machine are not manufactured to absolutely perfect specifications, they have always a certain eccentricity from the axis of the centre of rotation of the shaft. The eccentricity thus generates a centrifugal force. Thus, for example, for a shaft of 50 kN weight and an eccentricity of 0.1 mm, the centrifugal force generated is:

$$C = G/g \times e \times \omega^2 = 5000/9.81 \times 98800/10000 = 5035 \text{ kg}$$

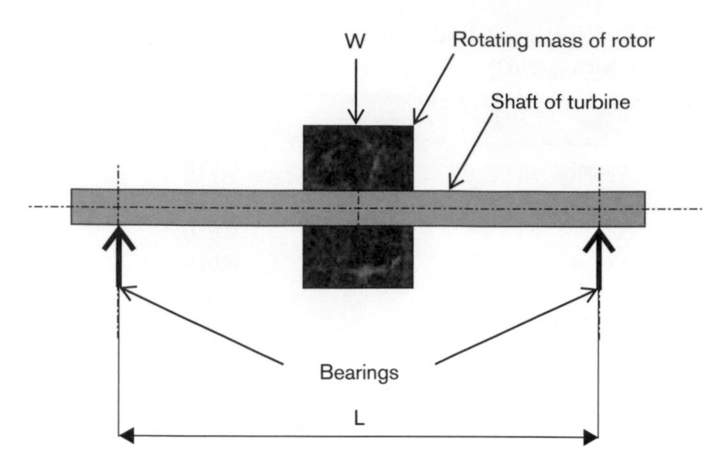

Fig. 10.4. Rotating mass of a rotor fixed to the shaft and simply supported on bearings

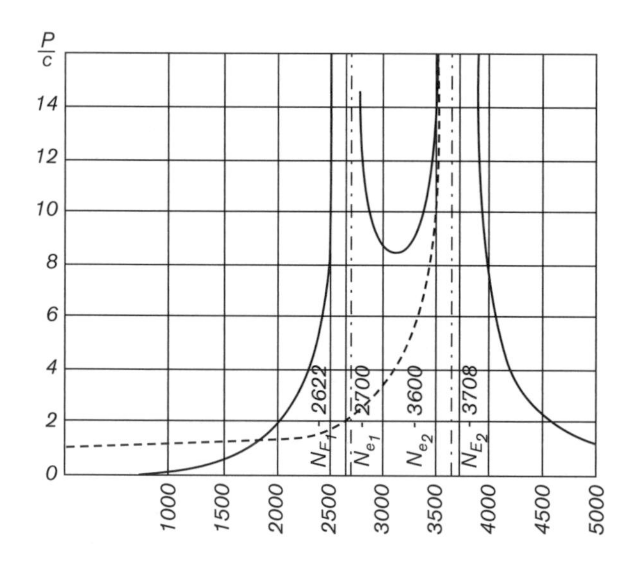

Fig. 10.5. Situation when the critical speed and the natural frequency of the supporting framework of the machine are too close to each other (after Fig. 287 by Major, 1962)

where $\omega^2 = [\pi \times N/30]^2 = 98800/\text{sec}^2$.

Dynamic Effects of high importance may occur when the critical speed and the natural frequency of the supporting framework of the turbo-generator machine are too close to one another, as shown in Fig. 10.5, where

N_{e1} = natural frequency of the turbine shaft (i.e. the critical speed) for a rigid support,
N_{e2} = natural frequency of the supporting framework of the machine (neglecting the soil elasticity),
N_{E1} and N_{E2} = the natural frequencies of the system with two degrees of freedom (the supporting structure of the machine and the turbine shaft with the rotor attached was regarded as a two-mass system).

Referring to Fig. 10.5, the results of the interaction between the critical speed and the natural frequency of the supporting framework lead us to make the following conclusions in order to prevent resonance occurring:

- There should be sufficient difference between the vertical natural vibrations (frequency) of the individual frames and the critical speed.
- Both the critical speed and the natural frequency of the supporting framework should differ by 20% from the operating speed of the machine.

10.6 Behaviour of material under cyclic loading (repetition of stress leading to fatigue)

The steel framework is subjected to an infinite number of repeated loads during the operation of the machine. The process of progressive localised permanent change occurring in this repeated loading produces fluctuating stresses and strains at some point or points, which may ultimately lead to complete fracture or failure of a member after a certain number of cycles. At this stage it may be defined as a *fatigue* failure of the material when it passes the endurance limit. Plastic deformation caused by dynamic fluctuating loads is of high importance.

The rate or frequency of cyclic loading causes no significant effect on the fatigue strength of steel when the applied stresses are relatively low and the frequency is less than 3000 c.p.m. But, if the stresses are high enough to produce plastic deformation with a cycle of loading, an increase in speed of loading will produce an increase in apparent fatigue strength. The magnitude of this effect can be determined only by tests.

An endurance test (fracture under a repeated load) on a specimen of structural steel was carried out by Murphy (1948). The specimen was a circular beam, turned down to a smaller diameter in the central portion. It was supported near the ends, and carried a weight so located that the highest stress would occur in the reduced section of the beam. Tensile stresses were developed on the lower half of the beam, and compressive stresses on the upper half. As the specimen was rotated by the motor connected to the beam at one end and the revolution counter connected at the other end, the magnitude of the longitudinal stress on any fibre varied through a complete cycle from maximum tension to maximum compression during each complete revolution. The set-up was arranged so that the motor stopped when the specimen broke; the revolution counter thus recorded the number of reversals of stress required to cause the failure. A number of tests on different specimens were carried out to obtain the best results.

The unit stresses of the specimen were plotted to a natural scale against the number of cycles (in logarithmic scale) required to cause failure. The graph thus produced is called an S–N (stress–cycle) diagram. See EC 3, Part 1-9, "Fatigue" (Eurocode, 2005). The endurance limit can be found from the S–N diagram as the unit stress at which the curve becomes horizontal. The diagram from the tests indicated the endurance limit in completely reversed bending to be 200 N/mm^2 (approximately) when the number of stress cycles reached beyond 10 million cycles, and there was a well-marked kink in the diagram. Thus, the endurance limit was lower than the elastic strength of the specimen, which was 250 N/mm^2.

Numerous tests on various specimens have shown that the endurance limit is approximately the same regardless of whether the stress is produced by bending or by axial loading. Therefore, in our case there is a rare possibility of the development of fatigue stress in the steel frame due to the machine running at 3000 r.p.m.

References

Eurocode, 3. BS EN 1993-1-9: 2005, Eurocode 3. Design of Steel Structures. Part 1-9: Fatigue.

Major, A., 1962. *Vibration Analysis and Design of Foundations for Machines and Turbines*, Collet's Holdings, London.

Murphy, G., 1948. *Properties of Engineering Materials*, International Textbook Company, Scranton, PA.

CHAPTER 11

Vibration Analysis

Before we start the structural analysis and design of the framework, the primary objective will be to carry out vibration analyses in order to verify that each transverse frame loaded with vertical forces has both vertical and horizontal frequencies (c.p.m.) differing in value from the normal operating speed (r.p.m.) and the critical speed of the machine by at least 20%, to avoid the occurrence of resonance of the structure. At the same time, the vertical and horizontal amplitudes of each frame should not exceed the maximum allowable limiting values as described in Chapter 10.

The elevated structural steel framework consists of three individual vertical transverse welded portal frames spaced at 4.375 m and 5.125 m from the left and connected by edge longitudinal beams. The frames are labelled *frame 1*, *frame 2* and *frame 3* counted from the left (see Figs. 9.1 and 9.2 in Chapter 9). Frame 1 supports half of the turbine weight, frame 2 supports half of the combined weight of the turbine and the generator, and frame 3 carries half of the generator weight. Vibration analyses will be performed for each individual frame to verify the stability of the frames against machine vibrations. We begin with a frequency analysis of the framework.

11.1 Determination of the vertical frequency

11.1.1 Vertical natural frequency of frame 1 (see Fig. 11.1a)

For the beam and column, try a section: UB 1016 × 305 × 314 kg/m + 2Flg.pls 600 × 30 × 141 kg/m, with

$A_s = 400 + 2 \times 60 \times 3 = 760\ \text{cm}^2 = 0.076\ \text{m}^2.$
$I_y = 644000 + 2 \times 60 \times 3 \times 51.5^2 = 1598810\ \text{cm}^4 = 0.016\ \text{m}^4.$
weight/m = 314 + 2 × 141 = 59/1 kg/m = 5.96 kN/m.
$E_s = 21 \times 10^7\ \text{kN/m}^2.$

and applying the deflection formulae of Kleinlogel (1958):

factor $k = I_b/I_c \times h/l = 5.71/4 = 1.43$

where

I_b = moment of inertia of beam = 0.016 m^4,
I_c = moment of inertia of column = 0.016 m^4,
h = height of column = 5.71 m,
l = span of beam = 4.0 m.

11.1.1.1 Loading

(a) UDL on beam = g_{k1} = 5.96 kN/m.
(b) Turbo-generator weight = W_{mach} = 235 kN at the centre of span of the beam.
(c) Vertical load on each column:

self-weight of column	= 5.96 × 5.71	= 34 kN
weight of longitudinal beam	= 9.59 × 2.19	= 21 kN
reaction from beam	= 5.96 × 2	= 12 kN

	Total	= 67 kN
add 10% for splay		= 6.7 kN

	Total	= 73.4 kN (74 kN, say)
load from turbo-generator	= 235/2	= 118 kN

	Total	= 192 kN.

11.1.1.2 Deflection

Deflection at mid span from W_{mach} and UDL:

$$\Delta_1 = 235 \times 4^3/(96 \times 21 \times 10^7 \times 0.016) \times (2 \times 1.43 + 1)/(1.43 + 3)$$
$$+ 5.96 \times 4^4/(384 \times 21 \times 10^7 \times 0.016) \times (5 \times 1.43 + 2)/(1.41 + 2)$$
$$= 4.38 \times 10^{-5} \text{ m.}$$

Deflection at the centre of the beam due to shearing forces:

$$\Delta_2 = 3/5 \times 4/(21 \times 10^7 \times 0.076) \times (235/2 + 5.96 \times 4/2) = 1.94 \times 10^{-5} \text{ m.}$$

Compression of the column due to vertical loads:

$$\Delta_3 = 5.71/(21 \times 10^7 \times 0.076) \times (192 + 235/2 + 5.96 \times 4/2) = 1.15 \times 10^{-4} \text{ m.}$$

The total deflection is:

$$\sum\Delta = 4.38 \times 10^{-5} + 1.94 \times 10^{-5} + 1.15 \times 10^{-4} = 1.78 \times 10^{-4} \text{ m.}$$

Therefore the vertical natural frequency of frame 1 is

$$N_{e1} = 30/(1.78 \times 10^{-4})^{0.5} = 2248 \text{ c.p.m.} < 3000 \text{ r.p.m. of the machine}$$

and

$$3000/2248 = 1.33 > 1.2 \text{ (minimum requirement).}$$

Satisfactory

11.1.2 Vertical natural frequency of frame 2 (see Fig. 11.1b)

For the beam and column, try a section: UB 1016 × 305 × 393 kg/m + 2Flg.pls 1000 × 30 × 236 kg/m, with

$A_s = 500 + 2 \times 100 \times 3 = 1100 \text{ cm}^2/100^2 = 0.11 \text{ m}^2$.
$I_y = 808000 + 2 \times 100 \times 3 \times 52.3^2 = 0.016 \text{ m}^4$.
weight/m = 393 + 2 × 236 = 8.65 kN/m.
$E_s = 21 \times 10^7 \text{ kN/m}^2$.
$k = 1.43$.

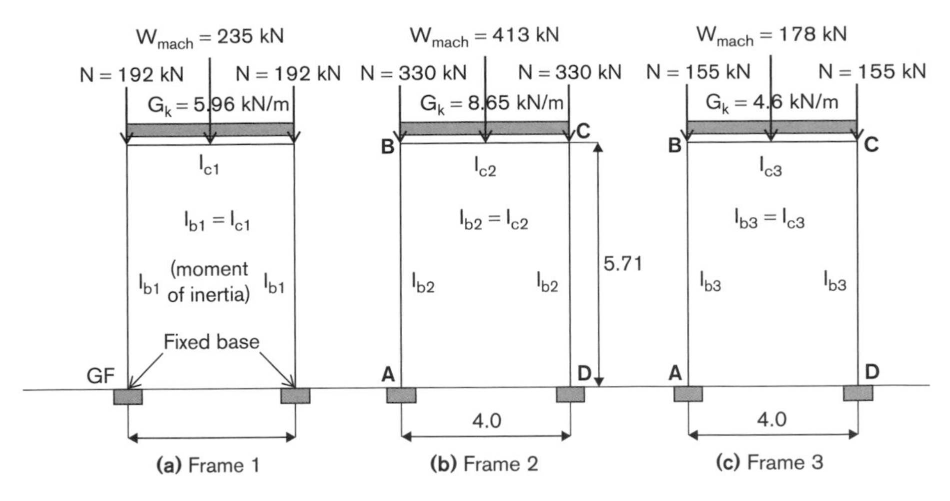

Fig. 11.1. Loadings on the three frames of an elevated steel structure to support a turbo-generator

11.1.2.1 Loading on transverse beam

(a) UDL on beam = g_{k2} = 8.65 kN/m.
(b) Turbo-generator weight = W_{mach} = 413 kN concentrated at mid span.
(c) Vertical load on each column:

self-weight of column = 8.65 × 5.71	= 49.4 kN
weight longitudinal beam	= (UB 1016 × 305 × 487 + 1000 × 30)
	= 9.6 × 4.75
	= 45.6 kN
reaction from beam = 8.65 × 2	= 17.3 kN

Total	= 112.3 kN
add 10% for splay	= 11.2 kN
weight of machine = 413/2	= 206.5 kN

Total	= 330 kN.

11.1.2.1 Deflection

Deflection at mid span due to machine load and UDL (g_{k2}):

$$\Delta_1 = W_{mach} \times l^3/(96 \times E_s \times 10^7 \times I_b) \times (2k + 1)/(k + 2) + (g_{k2} \times l^4)/(384 \times E_s \times I_b)/(5k + 2)/(k + 2)$$
$$= (413 \times 4^3)/(96 \times 21 \times 107 \times 0.016) \times (2 \times 1.43 + 1)/(1.43 + 2) + (8.65 \times 4^4)/(384 \times 21 \times 10^7 \times 0.016) \times (5 \times 1.43 + 2)/(1.43 + 2)$$
$$= 1.04 \times 10^{-4} \text{ m.}$$

Deflection at mid span due to shear forces by W_{mach} and g_{k2}:

$$\Delta_2 = 3/5 \times l/(E_s \times A_s)(W_{mach}/2 + g_{k2} \times l/2)$$
$$= 3/5 \times 4/(21 \times 10^7 \times 0.11) \times (413/2 + 8.65 \times 4/2) = 2.32 \times 10^{-5} \text{ m.}$$

Compression in the column due to vertical loads:

$$\Delta_3 = h/(E_s \times A_s)[N + (W_{mach} + g_{k2} \times l)/2]$$
$$= 5.71/(21 \times 10^7 \times 0.11) \times [330 + (413 + 8.65 \times 4)/2] = 1.37 \times 10^{-4} \text{ m.}$$

The total deflection is

$$\sum\Delta = (1.04 \times 10^{-4} + 2.32 \times 10^{-5} + 1.37 \times {}^{-4}) = 26.4 \times 10^{-5} \text{ m.}$$

Therefore the vertical natural frequency of frame 2 is:

$$N_{e2} = 30/(\Delta)^{0.5} = 30/(26.4^{-5})^{0.5} = 1846 \text{ c.p.m.} < 3000 \text{ r.p.m. of the machine}$$

and

$3000/1846 = 1.6 > 1.2$ (minimum requirement). Satisfactory

11.1.3 Vertical natural frequency of frame 3 (see Fig. 11.1c)

For the beam and column, try a section: UB 1016 × 305 × 222 kg/m + 2Flg.pls 500 × 30 × 118 kg/m, with

$A_s = 283 + 2 \times 50 \times 3 = 583 \text{ cm}^2 = 0.058 \text{ m}^2$.
$I_y = 40800 + 2 \times 5083 \times 50^2 = 1158000 \text{ cm}^4 = 0.011 \text{ m}^4$.
weight/m = 222 + 2 × 118 = 45/1 kg/m = 4.6 kN/m.
$E_s = 21 \times 1067 \text{ kN/m}^2$.
$k = 1.43$.

11.1.3.1 Loading

(a) UDL on beam = $g_k = 4.6$ kN/m.
(a) Generator weight = $W_{mach} = 255/2 = 178$ kN.
(a) Vertical load on each column:

self-weight of column	= 4.6 × 5.71	= 26 kN
weight longitudinal beam	= 9.59 × 2.56	= 24.6 kN
reaction from beam	= 4.6 × 2	= 9.2 kN

	Total	= 59.8 kN
add 10% for splay	= 5.9 kN	

	Total	= 65.7 kN
add reaction from generator		= 80.0 kN

	Total load on each column	= 155 kN.

11.1.3.2 Deflection

Deflection at mid span due to W_{mach} and UDL:

$$\Delta_1 = 178 \times 4^3/(96 \times 21 \times 10^7 \times 0.011) \times (2 \times 1.43 + 1)/(1.43 + 3)$$
$$+ 4.6 \times 4^4/(384 \times 21 \times 10^7 \times 0.011) \times (5 \times 1.43 + 2)/(1.43 + 2)$$
$$= 4.82 \times 10^{-5} \text{ m.}$$

Deflection due to shear forces by W_{mach} and UDL:

$$\Delta_2 = 3/5 \times 4/(21 \times 1067 \times 0.058) \times [178/2 + 4.6 \times 4/2] = 1.94 \times 10^{-5} \text{ m.}$$

Compression due to vertical loads on columns:

$\Delta_3 = 5.71/(21 \times 10^7 \times 0.058) \times [155 + 178/2 + 4.6 \times 4/2] = 1.19 \times 10^{-4}$ m.

The total deflection is:

$\sum\Delta = 18.64 \times 10^{-5}$ m.

Therefore the vertical natural frequency of frame 3 is:

$N_{e3} = 30/(18.64 \times 10^{-5})^{0.5} = 2197$ c.p.m. < 3000 r.p.m. of the machine

and

3000/2197 = 1.37 > 1.2 (minimum requirement). Satisfactory

11.1.4 Critical speed

As already stated, the natural frequencies of the frames must also differ from the critical speeds of the machine. The critical speed may be computed approximately from the single static deflection of the weight of the turbo-generator rotor. Thus,

total weight of turbo-generator = (350 + 235) kN.

Generally,

total weight of rotating mass = 18.6% of total weight of machine = 109 kN,
weight of rotor of turbine = 109 × 350/585 = 65 kN,
weight of rotor of generator = 109 × 235/585 = 44 kN.

Assume that the combined weight of the rotors of the turbine and generator shared

by frame 2 = G_{r2} = 65/2 + 44/2 = 55 kN;
by frame 1 = 65/2 = 32.5 kN;
by frame 3 = 44/2 = 22 kN.

Assuming that the shaft is simply supported, and that the load is concentrated at mid span of the beam of frame 2, we have

$\delta_c = G_r \times l^3/(6 \times E_s \times I_b) = 55 \times 4^3/(6 \times 21 \times 10^7 \times 0.016)$
= 2315 r.p.m. > 1959 c.p.m. of the frame
(δ_c = revolution of generator per minute)

and 2315/1959 = 1.18, which is only slightly less than 1.2 but can be accepted.

From the results of the above computations of the vertical natural frequencies of the three frames, we find that the value of the natural frequencies (c.p.m.) in the vertical planes are lower than the speed of the machine, which implies that the frames, being more flexible and of small mass, are undertuned.

It was mentioned earlier that following the results of measurements conducted by Major (1962) in his method for the synchronised measuring of vibrations, it was observed that at about 700 r.p.m. the amplitude has a maximum peak value as shown in Fig. 10.1, the natural frequency computed taking into account both the frame and the elasticity of foundation mass is found to be 700 c.p.m., and the peak of the amplitude is an indication of resonance. Similarly, at 1400 r.p.m., the amplitude is reasonably lower. Gradually, the fluctuation of amplitudes diminishes and a constant value is reached.

Thus, from the point of view of strength, the stresses due to maximum dynamic effect should be lower than the permissible stresses. For this reason, when $N_e < N_o$, which is the case of undertuning of the structure (the natural frequency of the structure is lower than the operating speed of the machine), the transient resonances occur during starting and stopping of the machine when the natural frequency of the structure may coincide with the speed of the machine. As a result, a dynamic force will develop and cause vibration. The transient resonances occur for a very short period and at speeds lower than normal operating speeds, with constant lower amplitudes of vibration. So, it does not appear to be dangerous to the structural integrity nor compromise safe performance of the structure.

So, for machines with speeds ranging from 1500 to 3000 r.p.m., the normal natural frequency of the framework shall be computed taking into consideration the bedding (foundation). As a check, the natural frequency of the foundation shall be computed taking into account the elasticity of concrete to verify the transient resonances.

11.1.5 Computation of the vertical natural frequency of the rigid foundation block, based on the elasticity of soil

11.1.5.1 Properties of soil under dynamic loads

Consider the foundation block 12.5 m long × 6.0 m wide × 2.5 m deep, placed on soil at a depth of 3.75 m. At 3.75 m below ground level, the soil consists of medium to dense clayey sand. From the results of field investigations and laboratory tests, the shearing strength of soil has been found to be 30°. The ground water table is assumed at ground level in the rainy season

The dynamic effect of vibrations and shocks of the machine tend to make the soil behave like a fluid and reduce the coefficient of internal friction of the soil considerably, by as much as 25 to 30% (Richart et al., 1998). Considering the properties of soil at a depth of 3.75 m that supports the foundation block, the coefficient of internal friction (Φ') should reduced. We assume a reduction factor of 30%. Therefore, the reduced value of Φ' is then 21°.

Referring to Terzaghi's ultimate bearing capacity equation:

ultimate bearing capacity $q_u = \gamma \times z \times N_q + 0.4 \times B \times \gamma \times N_y$

where N_q and N_y are bearing capacity factors, and with

$\Phi' = 21°$, $N_q = 8$, $N_y = 4.5$ (from Terzaghi's bearing capacity curves),
γ = density of soil submersed condition = (21 − 10) = 11 kN/m^3,
z = depth below ground level = 3.0 m,
B = width of foundation = 6 m.

we get

$$q_u = 11 \times 3 \times 8 + 0.4 \times 6 \times 11 \times 4.5 = 383 \text{ kN/m}^2$$

(the value of C_u is assumed to be zero).

C_u = undisturbed/undrained shear strength of soil (kN/m^2)

Assuming a factor of safety = 2.5, the dynamic safe bearing capacity of the soil is

$$q = 383/2.5 = 153 \text{ kN/m}^2.$$

(see Table 11.1).

Neglecting the deformation of the frame supporting the framework, and assuming a value of C_z = 35000 kN/m^3, the natural frequency of the foundation can be computed from the following equation:

$$N_s = 30/\Delta_s^{0.5}.$$

Table 11.1. Coefficient of uniform compression for various types of soil (based on Tables 12 and 13 by Major (1962))

Types of soil	Allowable stress q (kN/m^2)	Coefficient of uniform compression C_z (kN/m^3)
Peat, silt soils, loose sands, etc.	100 < 150	30000
Fine sands, medium to dense clayey sand saturated	100–150	60000
Medium-dense clayey sands saturated	150–250	100000
Course-grained gravelly medium to dense	> 250	> 100000

Here, Δ_s is the displacement of the foundation block and given by

$$\Delta_s = \sum\left[W_{mach} + W_{condenser} + G_{k1}\text{(framework)} + G_{k2}\text{(foundation block)}\right]/(B \times L \times C_z)$$

where

W_{mach} = total weight (turbine + generator) = $W_{tur} + W_{gene}$
= (470 + 355) kN = 825 kN,
$W_{condenser}$ = 165 kN,
G_{k1} = total weight of frames including longitudinal beams
= weight of (frame 1 + frame 2 + frame 3 + 2 × longitudinal beam)
= 91.9 + 133.4 + 70.93 + 182.2 = 478.4 kN
Adding 10% for splay
= 478.4 × 1.1 = 526 kN,
G_{k2} = weight of foundation block = 12.5 × 6 × 2.5 × 25 = 4688 kN,
$\sum G_k$ = (825 + 165 + 526 + 4688) kN = 6204 kN,

Therefore

$$\Delta_s = \sum G_k/(B \times L \times C_z) = 6204/(6.0 \times 12.5 \times 30000) = 2.4 \times 10^{-3}\ \text{m},$$

and so

$$N_s = 30/(2.4 \times 10^{-3})^{0.5} = 612\ \text{c.p.m.}$$

As already discussed before, referring to Fig. 10.1, when the elasticity of bedding (of the concrete foundation block) is considered to act simultaneously with the structural framework, the peak amplitude occurs at 700 r.p.m. and transient resonance is observed. In our case, the natural frequency computed above is 612 c.p.m., which is below the 700 r.p.m. of the machine. So, transient resonances are not likely to occur.

11.2 Computation of the horizontal frequency of the framework

11.2.1 Horizontal displacement due to a 1 kN load (assumed) acting horizontally along the axis of the frame

11.2.1.1 Consider frame 1

The total load carried by frame 1 is

G_{k1} = W_{mach} (weight of turbo-generator) + (self-weight of frame + weight of longitudinal beam)
= 235 + (91.9 + 42) × 1.1 = 382 kN.

Applying the deflection formulae of Kleinlogel (1958), the horizontal displacement due to a 1 kN load acting horizontally along the axis of the frame is

$$\Delta_{h1} = h^3/(12 \times E_s \times I_s) \times (3k + 2)/(6k + 1) + 6h/(5 \times E_s \times A_s) \times [1 + (A_s \times h)/(A_s \times L) \times 18k^2/(6k + 1)^2] + h^3/(E_s \times A_s \times L^2) \times 18k^2/(6k + 1)^2$$

where

h = height of column = 5.71 m,
L = span of frame = 4.0 m,
$E_s = 21 \times 10^7$ kN/m²,
A_s = cross-section of frame member = 0.076 m² (previously calculated),
I_s = moment of inertia of section = 0.016 m⁴ (previously calculated),
k = factor = 1.43 (previously calculated).

Therefore

$$\Delta_{h1} = 5.71^3/(12 \times 21 \times 10^7 \times 0.016) \times (3 \times 1.43 + 2)/(6 \times 1.43 + 1) + 6 \times 5.71/(5 \times 21 \times 10^7 \times 0.076) \times [1 + (0.076 \times 5.71)/(0.076 \times 4) \times 18 \times 1.43^2 /(6 \times 1.43 + 1)^2] + 5.71^3/(21 \times 10^7 \times 0.076 \times 4^2) \times 18 \times 1.43^2/(6 \times 1.43 + 1)^2$$
$$= 30.2 \times 10^{-7} + 6.75 \times 10^{-7} + 2.93 \times 10^{-7} = 39.9 \times 10^{-7} \text{ m}.$$

11.2.1.2 Consider frame 2

The total load carried by frame 2 is

G_{k2} = (self-weight of frame + weight of longitudinal frame) × 1.1 (for splay) + weight of machine
= (8.65 × 15.42 + 9.6 × 4.75) × 1.1 + 413 = 690 kN.

where

h = 5.71 m,
L = 4.0 m,
$E_s = 21 \times 10^7$ m²,
A_s = 0.11 m²,
I_s = 0.024 m⁴,
k = 1.43.

Therefore

$$\Delta_{h2} = 5.71^3/(12 \times 21 \times 10^7 \times 0.024) \times (3 \times 1.43 + 2)/(6 \times 1.43 + 1) + (6 \times 5.71)/(5 \times 21 \times 10^7 \times 0.11) \times [1 + (0.11 \times 5.71)/(0.11 \times 4) \times (18 \times 1.43^2/(6 \times 1.43 + 1)^2] + 5.71^3 /(21 \times 10^7 \times 0.11 \times 4^2) \times (18 \times 1.43^2)/(6 \times 1.43 + 1)^2$$
$$= 20.2 \times 10^{-7} + 4.7 \times 10^{-7} + 2 \times 10^{-7} = 26.9 \times 10^{-7} \text{ m}.$$

11.2.1.2 Consider frame 3

The total load carried by frame 3 is

G_{k3} = (self-weight of frame + weight of longitudinal beams) × 1.1 + (generator weight)/2
= (4.6 × 15.42 + 9.59 × 2.56) × 1.1 + 178 = 283 kN.

h = 5.71 m,
L = 4.0 m,
$E_s = 21 \times 10^7$ m²,

$A_s = 0.058$ m^2,
$I_s = 0.011$ m^4,
$k = 1.43$.

Therefore

$$\Delta_{h3} = 5.71^3/(12 \times 21 \times 10^7 \times 0.011) \times (3 \times 1.43 + 2)/(6 \times 1.43 + 1) + (6 \times 5.71)/ (5 \times 21 \times 10^7 \times 0.058) \times [1 + (0.058 \times 5.71)/(0.058 \times 4) \times (18 \times 1.43^2)/ (6 \times 1.43 + 1)^2] + 5.71^3/(21 \times 1067 \times 0.058 \times 4^2) \times (18 \times 1.43^2)/ (6 \times 1.43 + 1)^2$$
$$= 44 \times 10^{-7} + 8.8 \times 10^{-7} + 3.8^{-7} = 56.6^{-7} \text{ m}.$$

The factor of rigidity of the frame i.e. the force in kN causing a displacement of 1 m, is expressed by $H_i = 1/\Delta_i$ and so we have
∴

frame 1: $H_1 = 1/\Delta_1 = 1/(39.9 \times 10^{-7}) = 2.5 \times 10^5$ kN/m of displacement;
frame 2: $H_2 = 1/\Delta_2 = 1/(26.9 \times 10^{-7}) = 3.7 \times 10^5$ kN/m of displacement;
frame 3: $H_3 = 1/\Delta_3 = 1/(56.6 \times 10^{-7}) = 1.8 \times 10^5$ kN/m of displacement.

The total horizontal displacement due to a 1 kN load is equal to

$$\sum H = H_1 + H_2 + H_3 = (2.5 + 3.7 + 1.8) \times 10^5 = 8 \times 10^5 \text{ kN/m}.$$

11.2.2 To calculate the distance of the resultant vertical load from the axis of frame 1 (see Fig. 11.2)

Let X_g be the distance of the resultant load from the axis of frame 1. Taking the moment of all the vertical loads about the axis of frame 1, we get

$$\sum G_k \times X_g = G_{k2} \times 4.375 + G_{k3} \times 9.5,$$

i.e.

$$(382 + 690 + 283) \times x = 690 \times 4.375 + 283 \times 9.5 = 5707.$$

So

$$X_g = 5707/1355 = 4.2 \text{ m}.$$

11.2.3 To calculate the distance of the resultant force ∑H from the axis of frame 1

Let X_h be the distance of the resultant force from the axis of frame 1. Then

$$\sum H \times X_h = H_2 \times 4.375 + H_3 \times 9.5,$$

i.e.

$$8 \times 10^5 \times X_h = 3.7 \times 10^5 \times 4.375 + 1.8 \times 10^5 \times 9.5.$$

So

$$X_h = 33.3/8 = 4.1 \text{ m}.$$

Therefore, the eccentricity is

$$e = 4.2 - 4.1 = 0.1 \text{ m}$$

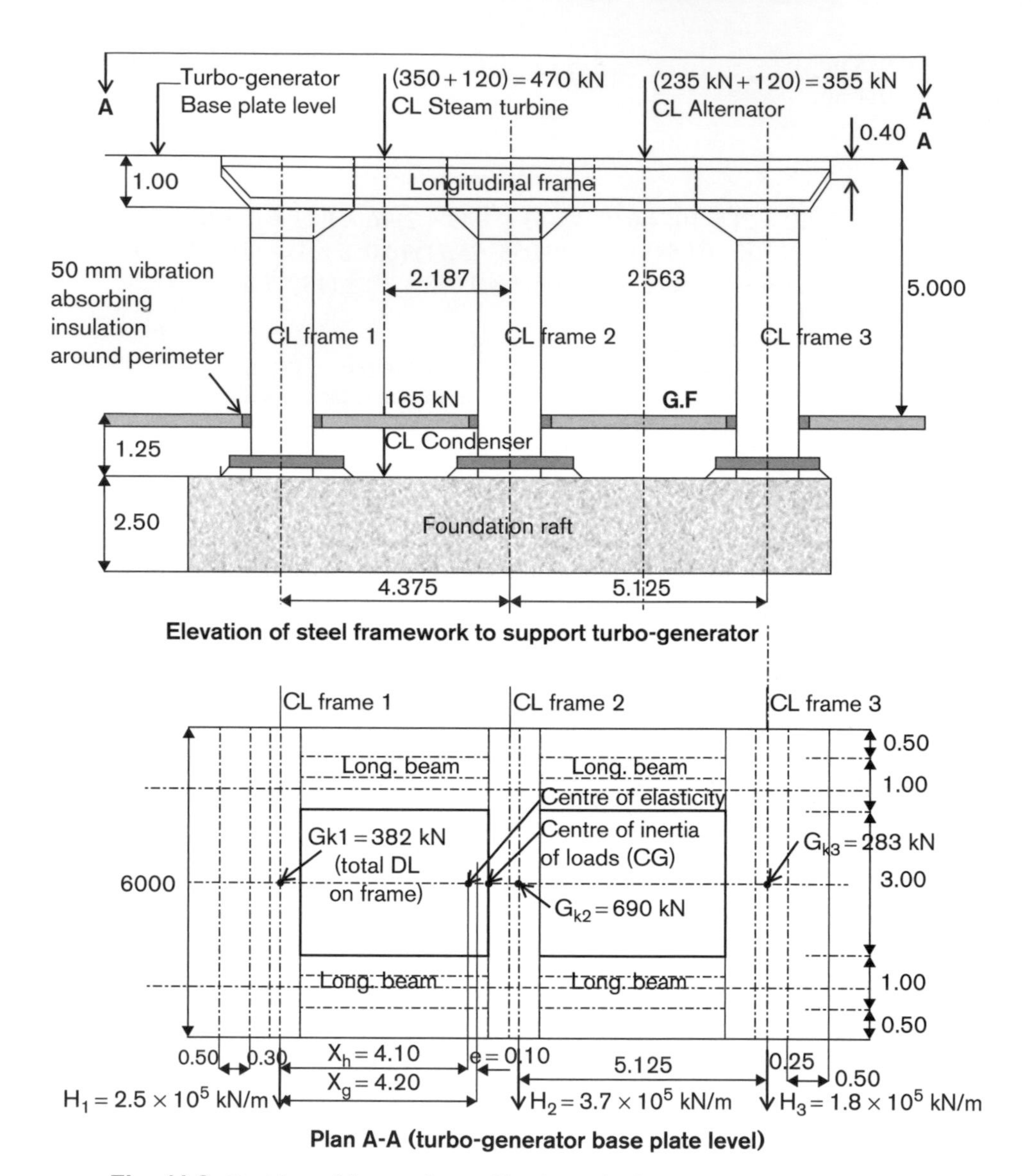

Fig. 11.2. Position of the resultant of loads on the frame and centre of elasticity

and

weight of frame 1 = G_{k1} = 382 kN;
weight of frame 2 = G_{k2} = 690 kN;
weight of frame 3 = G_{k3} = 283 kN.

11.2.4 To calculate the moment of inertia of the loading system about the resultant vertical load

$$
\begin{aligned}
I_g &= \sum G_i \times X_i^2 \text{ (see Fig. 11.2)} \\
&= G_{k1} \times X_g^2 + G_{k2} \times (4.375 - X_g)^2 + G_{k3} \times (9.5 - X_g)^2 \\
&= 382 \times 4.2^2 + 690 \times (4.375 - 4.2)^2 + 283 \times (9.5 - 4.2)^2 \\
&= 14709 \text{ kN m}^2.
\end{aligned}
$$

11.2.5 To calculate the moment of factor of rigidity of the frame about the resultant of the factor of rigidity

$$
\begin{aligned}
I_h &= \sum H_i \times X_h^{\,2} \\
&= H_1 \times X_h^{\,2} + H_2 \times (4.375 - X_h)^2 + H_3 \times (9.5 - X_h)^2 \\
&= 2.5 \times 10^5 \times 4.1^2 + 3.7 \times 10^5(4.375 - 4.1)^2 + 1.8 \times 10^5 \times (9.5 - 4.1)^2 \\
&= 94.8 \times 10^5 \text{ kN m}^2.
\end{aligned}
$$

11.2.6 Auxiliary value of Korchinski φ_0

$$
\begin{aligned}
\varphi_0 &= 1/2[e^2 \times \sum H/I_g + \sum H/\sum G_k + I_h/I_g] \\
&= 1/2[0.1^2 \times 8 \times 10^5/14709 + 8 \times 10^5/1355 + 94.8 \times 1065/14709] \\
&= 615.
\end{aligned}
$$

11.2.7 Horizontal natural frequency N_{eh}

The horizontal natural frequency can be expressed by equation 626 of Major (1962) as follows:

$$
\begin{aligned}
N_{eh} &= 30 \times (\Phi_0 \pm [\Phi_0^{\,2} \times (\sum H \times I_h)/(\sum G \times I_g)]^{0.5})^{0.5} \\
&= 30 \times (615 \pm [615^2 \times (8 \times 10^5 \times 94.8 \times 10^{-5})/(1355 \times 14709)]^{0.5})^{0.5} \\
&= 733 \text{ c.p.m. or } 735 \text{ c.p.m.} < 3000 \text{ r.p.m.} \qquad \underline{\text{Satisfactory}}
\end{aligned}
$$

11.3 Determination of vertical amplitudes

11.3.1 Frame 2 analysis

Frame 2 is the most heavily loaded frame subjected to the highest vibration amplitude. Therefore if this frame is of adequate strength, frames 1 and 3 are safe (lightly loaded), about half loaded. The vertical vibration amplitude may be given by the following expression:

$$A = \Delta_c \times \upsilon$$

where

υ = dynamic coefficient = $1/[(1 - N_o^{\,2}/N_{e2})^2 + (\Delta/\pi)^2 \times (N_o/N_e)^2]^{0.5}$,
Δ = logarithmic decrement of damping = 0.2 (see below),
N_o = speed of the machine = 3000 r.p.m.,
N_e = vertical natural frequency of the frame = 2004 c.p.m. (calculated previously).

Experimental results have shown that because of the damping effect of soil, the damping in turbine foundations is much higher than in the average reinforced concrete structure, for which it is only 0.25. When the soil effect is taken into account the value of the damping effect is taken equal to 0.4.

For a normal steel structure, experimental results have shown that, owing to its smaller mass, the value of the logarithmic decrement of damping is between 0.08 and 0.10. Experience has indicated that with the steel framework of a rigid welded portal construction for a turbine support, some damping occurs in the resonance zone during transition, owing to the friction at the interface between the machine base and its intermediate support. So, the value of Δ may be taken equal to 0.20. The dynamic coefficient is thus

$$\upsilon = 1/[(1 - 3000^2/2004^2)^2 + (0.2/3.14)^2 \times (3000/2004)^2]^{0.5} = 0.84.$$

The deflection Δ_c due to the centrifugal force at the centre of beam may be expressed in the following form (Kleinlogel, 1958):

$$
\begin{aligned}
\Delta_c = {} & C_e \times L^3/(96 \times E_s \times I_b) \times ((2k + 1)/(k + 1) + 3 \times C_e \times L/(5 \times E_s \times A_s) + \\
& h/(E_s \times A_s) \times C_e/2
\end{aligned}
$$

where

L = span of beam = 4.0 m,
h = height of column = 5.71 m,
E_s = elasticity modulus of steel = 21×10^7,
I_b = moment of inertia of section = 0.024 m^4,
A_s = cross-sectional area of beam = 0.11 m^2,
C_e = generating force due to the rotor movement with minimum eccentricity = $0.2 \times G_{r2}$,
G_{r2} = rotating mass of rotor connected to the shaft and supported on frame 2.

Now, to calculate the rotating mass of the rotors:

Total weight of turbo-generator = (350 + 235) kN = 585 kN,
Total weight of rotor (assumed) = 18.6% of total weight of turbo-generator
= 0.186×585 = 109 kN,
Weight of rotor of turbine = $109 \times 350/585$ = 65 kN,
Weight of rotor of generator = $109 \times 235/685$ = 44 kN,
Weight of rotors of turbine and generator is shared by the frame 2 (assumed)
= G_{r2} = (65 + 44)/2 = 54.5 (55 kN, say).

Since the natural frequency of frame 2 is less than the r.p.m. of the motor, i.e.

2204 c.p.m. < 3000 r.p.m. and not greater than 3000,

the theoretical value of the generating force $C_e = 0.2 \times G_{r2} = 0.2 \times 55 = 11$ kN. Therefore

$$\begin{aligned}\Delta_c &= 11 \times 4^3/(96 \times 21 \times 10^7 \times 0.024) \times (2 \times 1.43 + 1)/(1.43 + 2)\\ &\quad + (3 \times 11 \times 4)/(5 \times 21 \times 10^7 \times 0.11) + 5.71/(21 \times 10^7 \times 0.11) \times 11/2\\ &= 4.4 \times 10^{-6} \text{ m}.\end{aligned}$$

Therefore, the vertical amplitude of frame 2 is

$A = \Delta_c \times \upsilon = 4.4 \times 10^{-6} \times 0.84 \times 10^3 = 0.0037$ mm
<< permissible value 0.02 mm. Satisfactory

11.4 Determination of the horizontal vibration amplitude

Since the higher of the two horizontal natural frequencies (N_{eh2} = 753 c.p.m. or 735 c.p.m.) as already calculated is less than the operating speed of the machine (N_o = 3000 r.p.m.), the speed at the first peak of the resonance curve, i.e. during the acceleration and deceleration, is to be considered significant. Accordingly, it is assumed $N = N_{eh}$

The sum of the centrifugal forces is given by the expression:

$$\sum C_e = 0.2 \times \sum G_r \times (N_e/N_o)^2.$$

The weight of the rotating parts of the machine is

$$\sum G_r = 32 \text{ kN (frame 1)} + 55 \text{ kN (frame 2)} + 22 \text{ kN (frame 3)} = 109 \text{ kN}$$

and the sum of all centrifugal forces,

$$\sum C_e = 0.2 \times 109 \times (753/3000)^2 = 1.4 \text{ kN}.$$

The distance of the resultant of the centrifugal forces from the axis of frame 1 is equal to

$X_c = (55 \times 4.375 + 22 \times 9.5)/109 = 4.125$ m

and the distance of the resultant of the centrifugal forces from the axis of elasticity is equal to

$e_c = X_h - X_c = [-4.1$ (already calculated in the determination of horizontal natural frequencies) $+ 4.13] = -0.03$ m.

The centrifugal force acting upon the individual frame is given by the equation:

$C_i = \sum C_e \times H_i/\sum H_i + e_c \times \sum C_e \times H_i \times X_h/I_h$

where

$\sum H_i = 2.5 \times 10^5$ (for frame 1) $+ 3.7 \times 10^5$ (for frame 2) $+ 1.8 \times 10^5$ (for frame 3)
$= 8 \times 10^5$ kN/m,
$X_h = 4.1$ m (previously calculated in horizontal frequency determination),
$I_h = 94.8 \times 10^5$ (distance of frame from CG of resultant of factor of rigidity).

and

$C_1 = 1.4 \times 2.5 \times 10^5/(8 \times 10^5) + 0.03 \times 2.5 \times 10^5 \times 4.1/(94.8 \times 10^5) = 0.445$ kN;
$C_2 = 1.4 \times 3.7 \times 10^5/(8 \times 10^5) + 0.03 \times 1.4 \times 3.7 \times 10^5 \times 0.275/(94.8 \times 10^5) = 0.65$ kN;
$C_3 = 1.4 \times 1.8 \times 10^5/(8 \times 10^5) + 0.03 \times 1.4 \times 1.8 \times 1065 \times 5.225/(94.8 \times 10^5) = 0.314$ kN.

The horizontal displacement of the individual frames due to centrifugal force is therefore:

frame 1, $\delta_1 = C_1/H_1 = 0.445/(2.5 \times 10^5) = 1.78 \times 10^{-6}$ m;
frame 2, $\delta_2 = C_2/H_2 = = 0.65/(3.7 \times 10^5) = 1.76 \times 10^{-6}$ m;
frame 3, $\delta_3 = C_3/H_3 = 0.314/(1.8 \times 10^5) = 1.74 \times 10^{-6}$ m.

The horizontal natural frequencies of the frames are lower than the operating speed of the machine. So, the amplitudes shall be obtained by means of a maximum dynamic coefficient,

$\upsilon_{max} = \pi/\Delta = 3.14/0.2 = 15.7.$

Since the steel framework is fixed to the concrete foundation, we may assume a higher damping effect. So, we increase the logarithmic decrement of damping to 0.3, which results in the dynamic coefficient $\upsilon = 3.14/0.3 = 10.5$. The horizontal amplitudes are then

frame 1, $A_1 = \upsilon \times \delta_1 = 10.5 \times 1.78 \times 10^{-6} \times 10^3 = 0.018$ mm;
frame 2, $A_2 = \upsilon \times \delta_2 = 10.5 \times 1.76 \times 10^{-6} \times 10^3 = 0.018$ mm;
frame 3, $A_3 = \upsilon \times \delta_3 = 10.5 \times 1.74 \times 10^{-6} \times 10^3 = 0.018$ mm.

The maximum permissible amplitude is 0.03 mm, if less it is safer.

Satisfactory

See Figs. 11.3 and 11.4 for cross-sectional details of the framework.

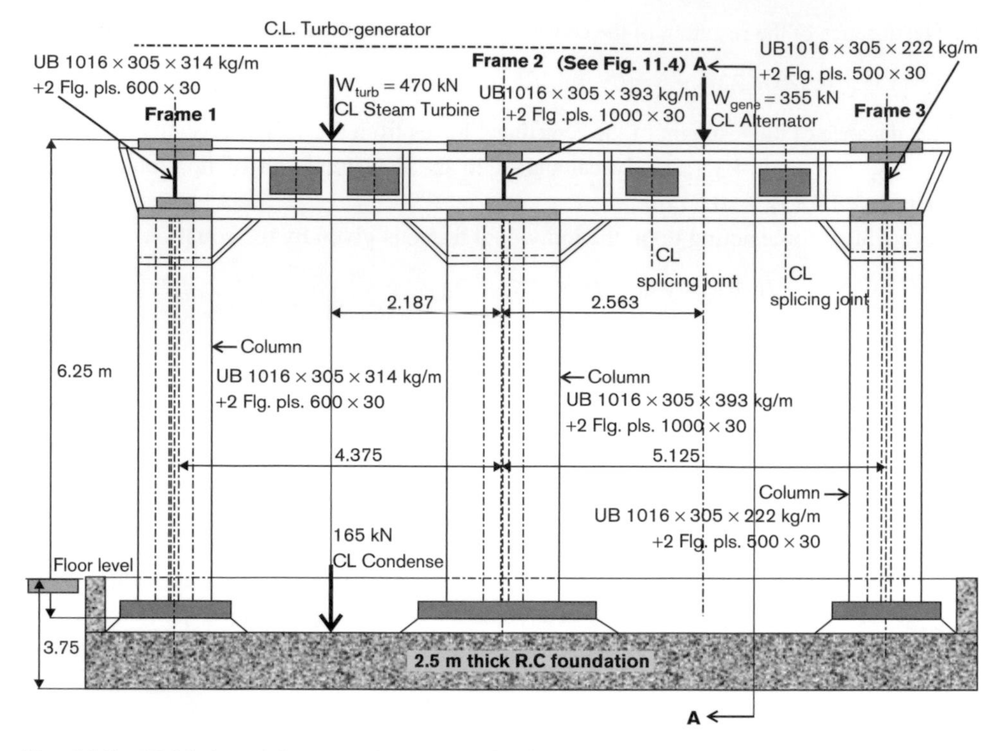

Fig. 11.3. Welded steel framework support of turbo-generator (showing cross-section of members)

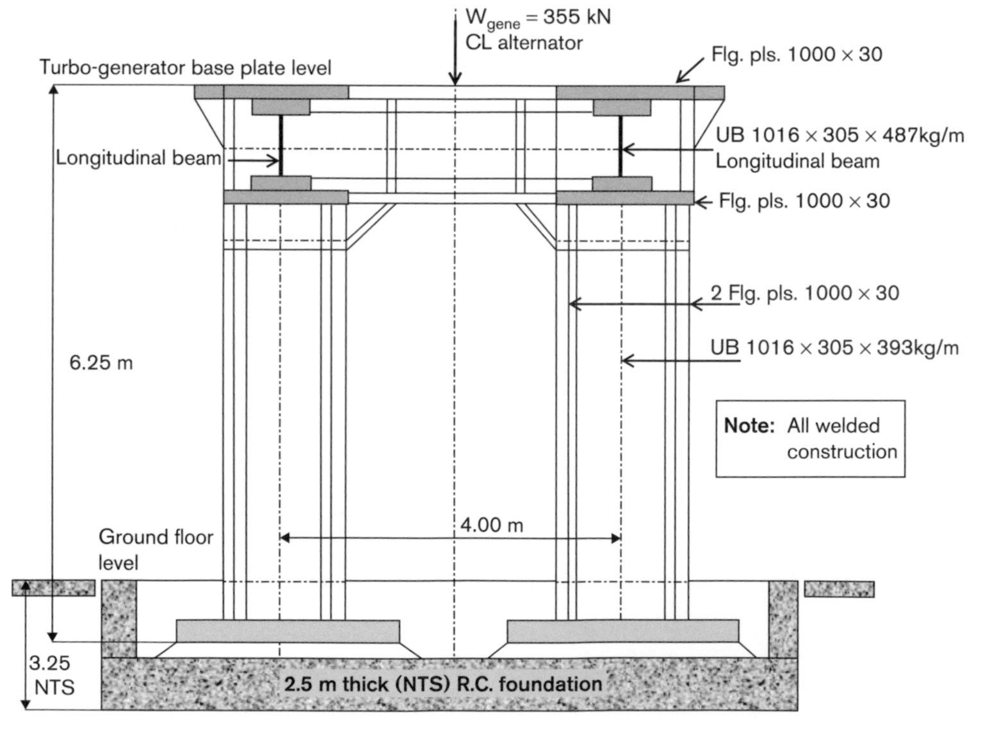

Fig. 11.4. Section A–A: Turbo-generator support frame 2

References

Kleinlogel, A.,1958. *Rahmenformeln "Rigid Frame Formulae"*. Wilhelm Ernst & Sohn, Berlin.

Major, A., 1962. *Vibration Analysis and Design of Foundations for Machines and Turbines*, Collet's Holdings, London.

Richart, F.E., Woods, R.D. and Hall, J.R., 1969. *Vibration of Soils and Foundations*, Prentice Hall Inc., N.J., U.S.A.

CHAPTER 12

Structural Analysis and Design

12.1 Structural analysis

Design considerations

The structural framework consists of three individual vertical transverse portal frames spaced at 4.375 m and 5.125 m from the left, connected by edge longitudinal beams, forming a base for the placement of a turbine and generator machine. The height of the frames from the reinforced concrete foundation raft to the top of the deck is 6.25 m. The transverse frames are subjected to machine dead loads and to vertical and horizontal dynamic forces, generated during the operation of the turbo-generator.

The frames will be analysed and the structural elements designed with the above force considerations. We will begin with frame 2.

12.1.1 Loadings on frame 2

12.1.1.1 Dead loads

(a) Self-weight of beam = 8.65 kN/m (previously calculated).
(b) Weight of machine shared by frame 2, including weight of rotor = 413 kN point load at mid span.

12.1.1.2 Dynamic forces

In addition to dead loads, the frame is subjected to vertical and horizontal dynamic forces that are generated due to the rotating parts of the turbo-generator. The dynamic force P may be expressed as

$$P = C \times \upsilon$$

where C is the centrifugal force corresponding to a bearing amplitude of 0.40 mm and υ is the dynamic factor.

Consider the three transverse frames that take the generated vertical and horizontal forces due to the rotating parts of the machine. The dynamic forces resulting from vibrations of the machine may create continuous fluctuating loads in the structural elements and thus may ultimately develop fatigue stresses. In this situation, a fatigue factor μ should be taken into account:

$$P_d = C \times \upsilon \times \mu.$$

A fatigue factor $\mu = 2.0$ should be assumed for a high-speed machine.

Relating the natural frequency of the structure to the speed of machine, the dynamic force may be computed as given below. If the natural frequency of machine is lower than the speed of the machine, i.e. $N_e \leq N_o$, then the maximum dynamic factor is

$$\upsilon_{max} = \pi/\Delta = 3.14/0.4 = 7.85 \text{ (8.0, say)}.$$

Thus

$$P_d = C \times 8 \times 2 = 16C.$$

When the speed of the machine is 3000 r.p.m., the centrifugal force caused by acceleration and deceleration is given by the following expression (equation 600 by Major (1962)):

$$C = 1.0G_{kr}(N_y/N_o)^2 \qquad (1)$$

where

G_{kr} = weight of rotating parts of the machine,
N_y = natural frequencies of the frame, corrected by a factor $(1 \pm a_e)$.

Therefore, the dynamic force is

$$P_d = 16 \times G_{kr}(N_y/N_o)^2.$$

The values of the frequencies obtained are subjected to an error of ±5–15% due to the simplification of the transverse frame calculations, in neglecting the effect of the longitudinal beams and foundation bedding mass.

So, in calculating the dynamic forces, the natural frequencies are adjusted by a correction factor a_e of 0.1–0.3. We may assume an average value for the correction factor, i.e. $a_e = 0.20$, as it is difficult to access accurately all the factors influencing the natural frequencies.

Vertical dynamic forces

Since the natural frequencies of frames 1, 2, and 3 ((2248, 1846 and 2197 + $(1 \pm a_e)$, respectively) are lower than 3000 r.p.m., and assuming $N_y = N_o$, the maximum dynamic force is given by

$$P_{max} = 16 \times G_{kr}.$$

Now, the weight of rotating parts of the turbine and generator shared by each frame is

frame 1, $G_{kr1} = 32.5$ kN;
frame 2, $G_{kr2} = 55$ kN;
frame 3, $G_{kr3} = 22$ kN;

(these values have been previously calculated). Therefore the vertical dynamic force acting on each frame is

frame 1, $P_{v1} = 16 \times 32.5 = 520$ kN;
frame 2, $P_{v2} = 16 \times 55 = 880$ kN;
frame 3, $P_{v3} = 16 \times 22 = 352$ kN;

(see Fig. 12.1).

Horizontal dynamic forces

Horizontal natural frequency $N_{eh1} = 733$ c.p.m. and $N_{eh2} = 735$ c.p.m.
Corrected higher natural frequency $N_y = N_{eh2} \times (1 + a_e) = 735 \times 1.2 = 882$ c.p.m.
Corrected lower r.p.m. of the machine $= N_o/(1 + a_e) = 3000/1.2 = 2500$ r.p.m.

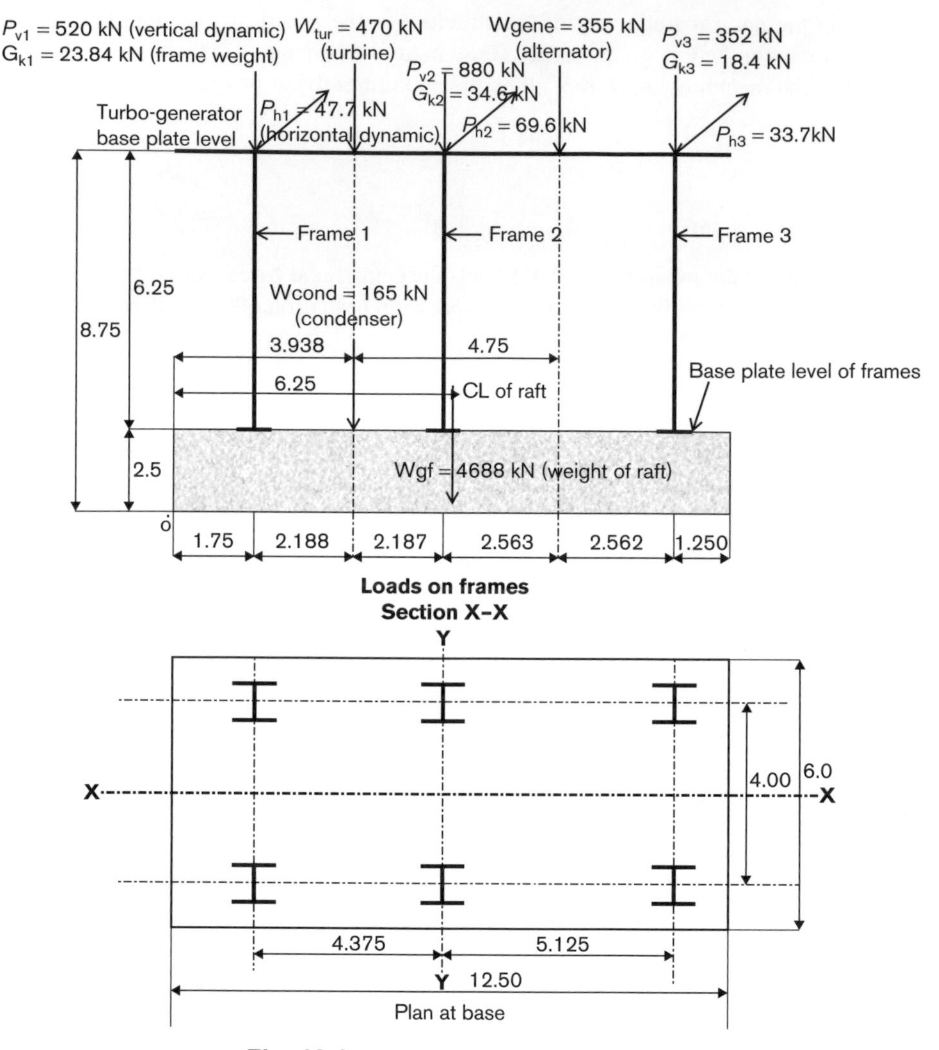

Fig. 12.1. Loads on frames at deck level

Since the higher natural frequency is lower than the corrected r.p.m. of the machine, the total horizontal dynamic force is

$$P_h = 16\sum G_{kr} \times [N_y/N_o]^2$$

where

$$\sum G_{kr} = G_{kr1} + G_{kr2} + G_{kr3} = 32.5 + 55 + 22 = 109.5 \text{ kN.}$$

So

$$P_h = 16 \times 109.5 \times [882/3000]^2 = 151 \text{ kN.}$$

The dynamic force acting upon an individual frame i is

$$P_{hi} = P_h \times C_i/\sum C_i$$

where

C_i = centrifugal force acting upon an individual frame i,

C_1 = 0.445 kN acting upon frame 1,
C_2 = 0.65 kN acting upon frame 2,
C_3 = 0.314 kN acting upon frame 3,
$\sum C_i = (0.445 + 0.65 + 0.314)$ kN = 1.41 kN.

Therefore, the dynamic horizontal force acting on each frame is

frame 1, $P_{h1} = 151 \times 0.445/1.41 = 47.7$ kN;
frame 2, $P_{h2} = 151 \times 0.65/1.41 = 69.6$ kN;
frame 3, $P_{h3} = 151 \times 0.314/1.41 = 33.7$ kN;

(see Fig. 12.1).

12.1.2 Structural analysis of frame 2

12.1.2.1 Structural model

Prepare a structural model with beam and columns, having a base that is assumed fixed at the foundation raft. The beam and column connections is assumed fixed (continuous), as are the bases of the columns. Consider the frame labelled as "abcd" along the centre line (CL) of the frame (see Fig. 12.2).

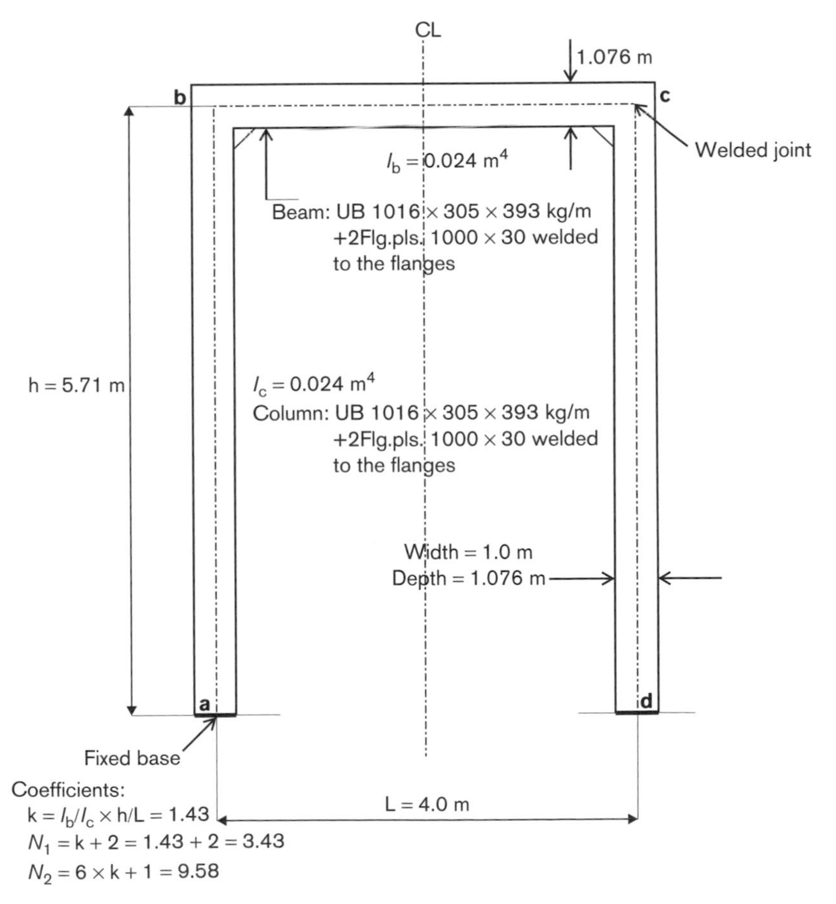

Fig. 12.2. Structural model for frame 2

12.1.2.2 Assumed sections, properties and coefficients

Beam: UB 1016 × 305 × 393 kg/m + 2Flg.pls. 1000 × 30 mm.
$A = 0.11$ m²; $I_{by} = 0.024$ m⁴.
L = effective length of the beam reckoned along the centre line of the frame = 4.0 m.
Column: UB 1016 × 305 × 39/1 kg/m + 2Flg.pls. 1000 × 30 mm.
$A = 0.11$ m²; $I_{cy} = 0.024$ m⁴.
h = effective height of the column reckoned along the centre line of the frame = 5.71 m.

Using well-established formulae from Kleinlogel (1958), we can determine the coefficients:

$k = I_{by}/I_{cy} \times h/L = 0.024/0.024 \times 5.71/4.0 = 1.43.$
$N_1 = k + 2 = 1.43 + 2 = 3.43.$
$N_2 = 6k + 1 = 6 \times 1.43 + 1 = 9.58.$

12.1.2.3 Moments

Due to uniform dead loads on the beam

$g_k = 8.65$ kN/m.
$M_{ab} = -M_{dc} = g_k \times L^2/(12 \times N_1) = 8.65 \times 4^2/(12 \times 3.43) = 3.4$ kN m.
$-M_{bc} = +M_{cb} = -g_k \times L^2/(6N_1) = -8.65 \times 4^2/(6 \times 3.43) = -6.7$ kN m.
M_{max} mid bc $= g_k \times L^2/8 + (M_{bc}) = 8.65 \times 4^2/8 - 6.7 = 10.6$ kN m.
$V_a = V_d = g_k \times L/2 = 8.65 \times 4/2 = 17.3$ kN; $H_a = -H_d = 3 \times M_a/h = 3 \times 3.4/5.71 = 1.8$ kN.(See Fig. 12.3a.)

Due to machine load on the beam

$W_{mach} = 413$ kN acting at mid span of the beam.
$M_{ab} = -M_{dc} = W_{mach} \times L/(8 \times N_1) = 413 \times 4/(8 \times 3.43) = 60.2$ kN m.
$M_{bc} = M_{cb} = -2 \times M_{ab} = -2 \times 60.2 = -120.4$ kN m.
$V_a = W_{mach}/2 = 413/2 = 206.5$ kN.
$H_a = 3 \times M_{ab}/h = 3 \times 60.2/5.71 = 31.6$ kN.
M_{mid} bc $= W_{mach} \times L/4 - M_{bc} = 413 \times 4/4 - 120.4 = 292.6$ kN. (See Fig. 12.3b.)

(M_{max} mid bc = maximum moment at middle of span of beam bc; M_{mid} bc = moment at middle of span of beam bc)

Due to dynamic vertical force on the beam

$P_{d2} = 880$ kN acting at mid span.
$M_{ab} = -M_{dc} = P_{v2} \times L/(8N_1) = 880 \times 4/(8 \times 3.43) = 128.3$ kN m.
$-M_{bc} = M_{cb} = 2 \times M_{ab} = 256.6$ kN m.
$V_a = V_d = P_{v2}/2 = 880/2 = 440$ kN.
$H_a = -H_d = 3 \times M_{ab}/h = 3 \times 128.3/5.71 = 67.4$ kN.
M_{mid} bc $= P_{v2} \times L/4 - M_{bc} = 880 \times 4/4 - 256.6 = 623.4$ kN m. (See Fig. 12.4a.)

Due to dynamic horizontal force on the beam

$P_{h2} = 69.6$ kN acting at beam level.
$-M_{ab} = -M_{dc} = P_{h2} \times h/2 \times (3 \times k + 1)/N_2$
$= -[69.6 \times 5.71/2 \times (3 \times 1.43 + 1)/9.58] = -109.7$ kN m.

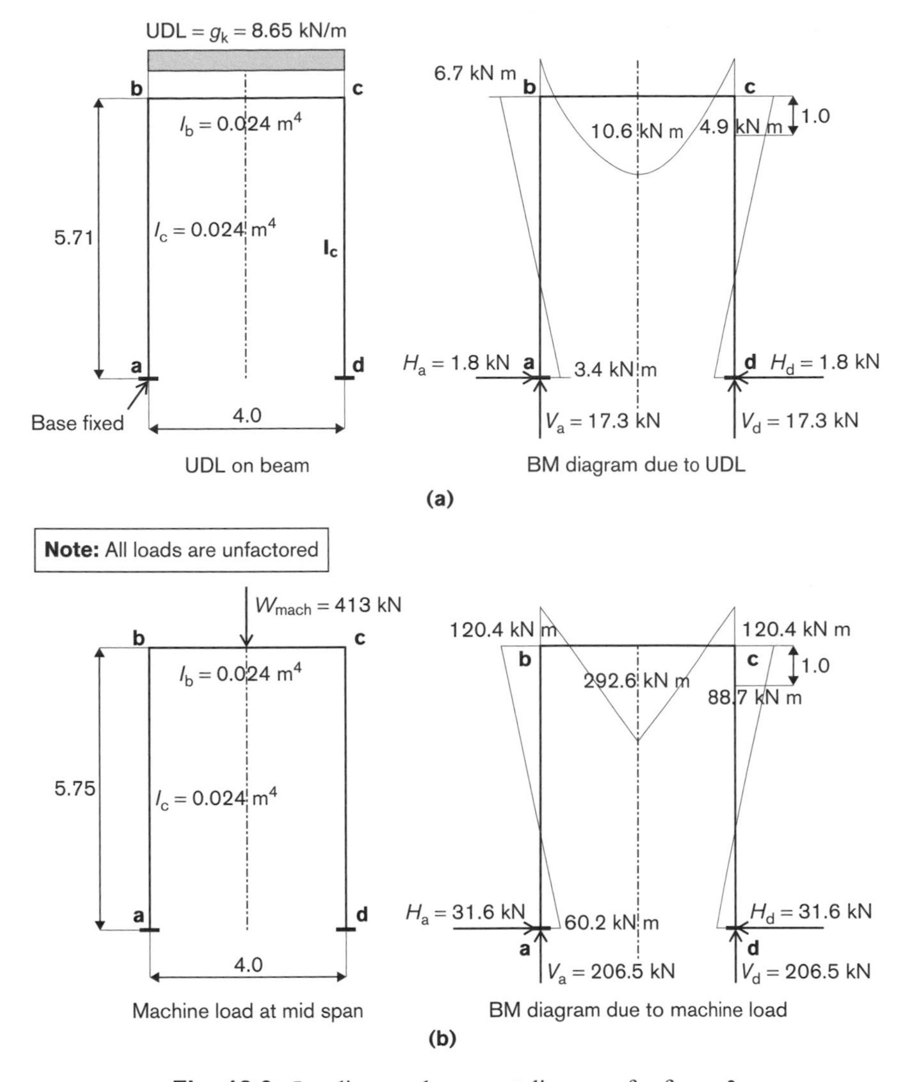

Fig. 12.3. Loadings and moment diagrams for frame 2

$$-M_{ba} = M_{bc} = -[P_{h2} \times h/2 \times (3 \times k)/N_2] = -[69.6 \times 5.71/2 \times (3 \times 1.43)/9.58]$$
$$= -89.0 \text{ kN m.}$$

$$-H_a = -H_d = P_{h2}/2 = 69.6/2 = -34.8 \text{ kN.}$$
$$-V_a = V_d = 2 \times M_{ba}/L = 2 \times 109.7/4 = -54.9 \text{ kN.} \qquad \text{(See Fig. 12.4b.)}$$

Note: the formulae are derived from Kleinlogel (1958).

12.1.2.4 Design load combinations (ULS method)

Referring to Clause 6.4.3.2, "Combinations of actions for persistent or transient design situations (fundamental combinations)", of BS EN 1990: 2002(E) (Eurocode, 2002), the load combination is given by the expression:

$$\Sigma\gamma_{G,j} \times G_{k,j} \text{ “+” } \gamma \times P \text{ “+” } \gamma_{Q,1} \times Q_{k,1} \text{ “+” } \Sigma\gamma_{Q,i} \times \psi_{0,i} \times Q_{k,i} \qquad (2)$$

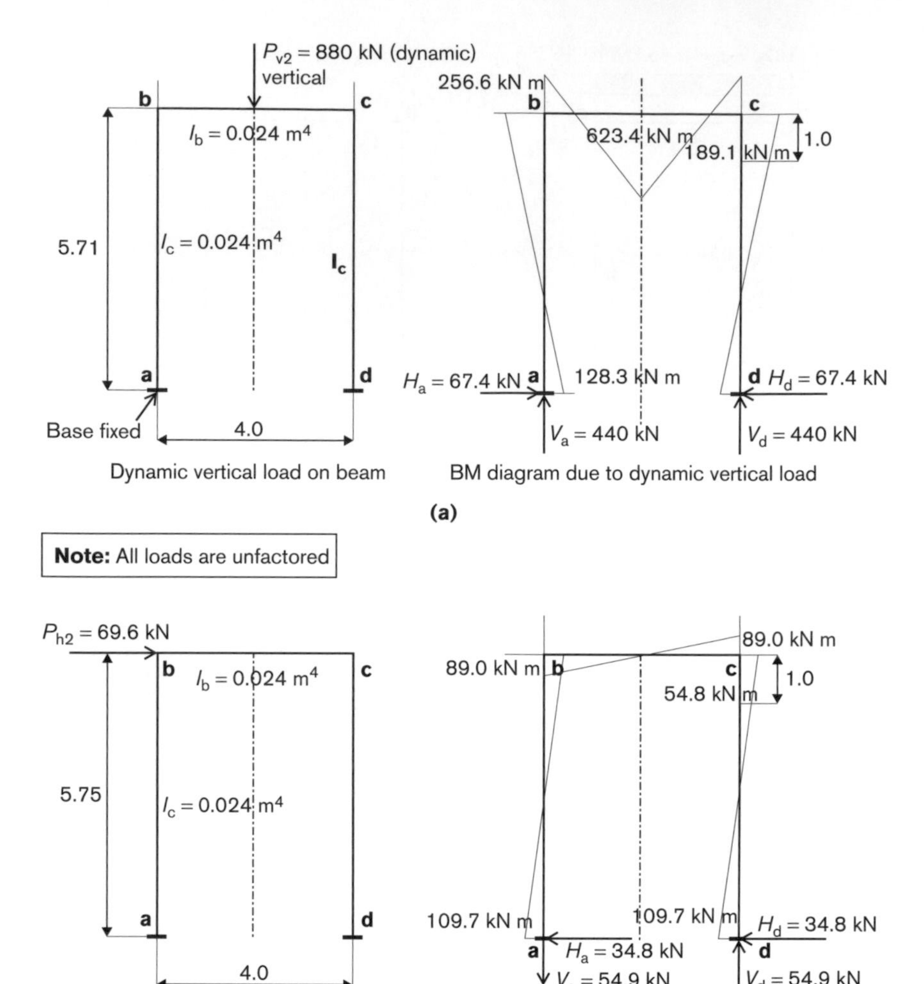

Fig. 12.4. Vertical and horizontal dynamic loadings and moment diagrams for frame 2

where

$\gamma_{G,j}$ = partial factor for permanent action j,
$G_{k,j}$ = characteristic value of permanent action j,
P = relevant representative value of a prestressing action,
$\gamma_{Q,1}$ = partial factor for variable action 1,
$Q_{k,1}$ = characteristic value of leading variable action 1,
$\gamma_{Q,i}$ = partial factor for variable action i,
$\psi_{0,i}$ = factor for combination value of a variable action,
$Q_{k,i}$ = characteristic value of accompanying variable action i.

In our case, no prestressing P action is applied. So, the above expression (2) is reduced to

$$\Sigma\gamma_{G,j} \times G_{k,j} + \gamma_{Q,1} \times Q_{k,1} + \Sigma\gamma_{Q,i} \times \psi_{0,i} \times Q_{k,i}.$$

12.2 Design of section of structural elements

- The design of sections shall be based on EC 3, Part 1, 2005 except where stated otherwise (Eurocode, 2005).
- In the following, all Clauses referred to shall imply the above code unless mentioned otherwise.
- The design of section will be carried out by the ultimate-limit-state method.

12.2.1 Beam

Consider the section at mid span.

12.2.1.1 Characteristic moment

From final results already calculated for moments and forces:

Moment at mid span due to UDL $M_{gk} = 10.6$ kN m (permanent actions).
Moment at mid span due to machine load $M_{Gk} = 262.6$ kN m (permanent actions).
Moment at mid span due to vertical dynamic loading $M_{Qk} = 623.4$ kN m (variable actions).

12.2.1.2 Ultimate design values

In combinations of loadings in the ultimate-limit-state design method, the partial factors shall be multiplied by the characteristic values. Referring to Table A1.2(A), "Design values of actions (EQU)(Set A)", of values of γ for the ULS method:

$\Sigma\gamma_{G,j} \times G_{k,j} + \gamma_{Q,1} \times Q_{k,1} + \gamma_{Q,i} \times \psi_{0,i} \times Q_{k,i}$ [with accompanying variable (horizontal dynamic) load]

where the partial factors are

for combined (UDL + machine dead load): $\gamma_{Gj} = 1.35$ (permanent actions)
for vertical dynamic loadings: $\gamma_{Q,1} = 1.5$ (leading variable action)
for horizontal dynamic loadings: $\gamma_{Q,i} = 1.5$

and $\psi = 0.7$ (see Table A1.1 BS EN 1990: 2002(E)), the maximum ultimate design moment is therefore

$$
\begin{aligned}
M_{Ed} &= \gamma_{Gj} \times (M_{gk} + M_{Gk}) + \gamma_{Q,1} \times M_{Qk} \\
&= 1.35 \times 273.2 + 1.5 \times 623.4 \\
&= 1304 \text{ kN m},
\end{aligned}
$$

and the maximum ultimate design shear at the end of splay is

$$V_{Ed} = 1.35 \times (8.65 + 206.5) + 1.5 \times 440 + 1.5 \times 0.7 \times 54.9 = 1008 \text{ kN}.$$

So, the maximum ultimate thrust is

$N_{Ed} = 1.35 \times$ [shear at column base due to (UDL + weight of machine)]
$+ 1.5 \times$ [shear at column base due to dynamic vertical $+ 1.5 \times 0.7$ (horizontal dynamic loads)]
$= 1.35 \times (1.8 + 31.6) + 1.5 \times (67.4) + 1.5 \times 0.7 \times 34.8 = 183$ kN.

12.2.1.3 Initial sizing of section

Try the section: UB 1016 × 305 × 393 kg/m + 2Flg.pls 1000 × 30.

Properties of section

Depth of section h = 1076 mm.
Depth between fillets d = 956 mm.
Width of section b = 1000 mm.
Thickness of the web t_w = 24.4 mm.
Average thickness of the flange t_f = 37 mm.
Root radius r = 30 mm.
Area of section A = 1100 cm^2.
Moment of inertia I_y = 2449174 cm^4.
Moment of inertia I_z = 520500 cm^4.
Section modulus W_y = 45523 cm^3.
Section modulus W_z = 10410 cm^3.
Radius of gyration i_y = $\sqrt{(I_y/A)}$ = 47.2 cm.
Radius of gyration i_z = $\sqrt{(I_z/A)}$ = 21.8 cm.

12.2.1.4 Section classification

For the following, we refer to Table 5.2 of EC 3: Part 1-1 (see Appendix).

Flange

Assumed,

f_y = 275 N/mm^2 for steel grade S 275.
Stress factor $\varepsilon = \sqrt{(235/275)} = 0.92$ $t \leq 40$ mm.
c = 1000/2 − 303/2 = 348.5 mm.
Ratio c/t_f = 348.5/30 = 11.6.
$9\varepsilon/\alpha$ = 9 × 0.92 = 8.3.

For class 1 classification:

limiting value of $c/t_f \leq 9\varepsilon$.

In our case, c/t_f (11.6) > $9\varepsilon/\alpha$ (8.3), so the condition for class 1 section classification is not satisfied. Therefore, stiffeners shall be provided to prevent buckling of the flange plate.

Web

Ratio d/t_w = 956/24.4 = 39.2.

As the web is subjected to bending and compression, and assuming the depth of compression is less than half the depth of the section, i.e. $\alpha \leq 0.5$, we have

d/t_w = 39.2;
$36 \times \varepsilon/\alpha = 36 \times 0.92/0.5 = 66.3$.

For class 1 section classification:

limiting value of $d/t_w \leq 36 \times \varepsilon/\alpha$.
In our case, d/t_w (39.2) < $36 \times \varepsilon/\alpha$ (66.3). Therefore, the condition is satisfied.

12.2.1.5 Moment capacity

Maximum ultimate design moment M_{Ed} = 1304 kN m.
Maximum ultimate shear V_{Ed} = 1008 kN.
Maximum ultimate thrust N_{Ed} = 183 kN.

Referring to Clause 6.2.10 of EC 3, where bending moment, shear and axial thrust act simultaneously on a structural member, the moment capacity of the section shall be calculated in the following way:

- Where shear and axial force are present, allowance shall be made for the effect of both shear force and axial thrust on the resistance moment.
- Provided that the design value of the shear force V_{Ed} does not exceed 50% of the plastic shear resistance $V_{pl,Rd}$, no reduction of the resistance defined for the bending and axial force in Clause 6.2.9 need be made, except where shear buckling reduces the section resistance.
- Where V_{Ed} exceeds 50% of $V_{pl,Rd}$, the design resistance of the cross-section to the combination of moment and axial force should be calculated using a reduced yield strength $(1 - \rho) \times f_y$ of the shear area, where $\rho = (2V_{Ed}/V_{pl,Rd} - 1)^2$.

(a) *When the web is not susceptible to buckling.*
When the web depth to thickness ratio $d/t_w \leq 36 \times \varepsilon/\alpha$ for class 1 section classification, it should be assumed that the web is not susceptible to buckling, and the moment capacity shall be calculated by the expression $M_{Rd} = f_y \times W_y$. In our case we have assumed a section in which $d/t_f < 36 \times \varepsilon/\alpha$.

(b) *When the ultimate shear force $V_{Ed} \leq 0.5V_{pl,Rd}$.*
The ultimate shear force at the end of splay $V_{Ed} = 1008$ kN and the design plastic shear capacity of the section is

$$V_{pl,Rd} = A_v \times (f_y/\sqrt{3}/\gamma_{Mo})$$

where

$$A_v = \text{shear area} = A - 2 \times b \times t_f + 2(t_w + 2r) \times t_f$$
$$= 1100 - 2 \times 100 \times 3.7 + 2 \times (2.44 + 2 \times 3) \times 3.7$$
$$= 422 \text{ cm}^2,$$
$$\gamma_{Mo} = 1.0.$$

Therefore

$$V_{pl,Rd} = 422 \times 100 \times (275/\sqrt{3})/1000 = 6700 \text{ kN},$$

and

$$0.5 \times V_{pl,Rd} = 0.5 \times 6700 = 3350 \text{ kN}.$$

So,

$$V_{Ed}\ (1008 \text{ kN}) < 0.5 \times V_{pl,Rd}\ (3350 \text{ kN}).$$

Therefore, the effect of shear force on the reduction of the plastic resistance moment need not be considered.

(c) *When the ultimate axial force $N_{Ed} \leq 0.25N_{pl,Rd}$*
But, $0.25 \times N_{pl,Rd} = A \times f_y/\gamma_{Mo} = 0.25 \times 1100 \times 10^2 \times 275/10^3$ $= 7563$ kN. In our case, N_{Ed} (183 kN) << $0.25 \times N_{pl,Rd}$ (7563 kN). So, the effect of axial force on the reduction of the plastic resistance moment need not be considered and the plastic moment capacity is

$$M_{pl,Rd} = f_y \times W_{ply} = 275 \times 45523 \times 10^3/10^6$$
$$= 12518 \text{ kN m} >> N_{Ed}\ (1304 \text{ kN m}).$$

<u>Satisfactory</u>

12.2.1.6 Shear buckling resistance

Shear buckling resistance need not be checked if the ratio $h_w/t_w \leq 72\varepsilon$. But

$d/t_w = 39.2$ (previously calculated)

and

$36 \times \varepsilon/\alpha = = 66.3.$

So,

$d/t_w\ (39.2) < 36 \times \varepsilon/\alpha\ (66.3).$

Therefore, shear buckling resistance need not be checked.

12.2.1.7 Buckling resistance moment

The top compression flange is susceptible to torsional buckling. The end of the beam is assumed fixed. So the length of unrestrained compression flange may be taken to be equal to 0.75×4.0 (centre to centre of beam) = 3.0 m.

Referring to Clause 6.3.2.2, the value of χ_{LT} (reduction factor for lateral torsional buckling) is computed by first determining the appropriate non-dimensional slenderness,

$\bar{\lambda}_{LT} = \sqrt{[(W_y \times f_y)/M_{cr}]}$

where M_{cr} = elastic critical moment for lateral torsional buckling. To determine the value of M_{cr}, we have to go through complex equations and time-consuming calculations. For practical purposes, the buckling resistance moment is calculated in the following way.

Referring to Clause 6.3.2.4 ("Simplified assessment methods for beams with restraints in buildings"), the lateral restraint of the length to the compression flange of a member will prevent any lateral torsional buckling if it satisfies the following condition:

$$\bar{\lambda}_f = k_c \times L_c/(i_{f,z} \times \lambda_1) \leq \bar{\lambda}_{c0} \times (M_{c,Rd})/M_{y,Ed} \qquad (12.3)$$

where

k_c = slenderness correction factor = 0.90 assuming both ends fixed (see Table 6.6 of EC 3),
L_c = restraint length = 300 cm.

If,

z = radius of gyration of the equivalent compression flange composed of flange
$= \sqrt{(I_z/A)} = \sqrt{(260200/433.3} = 24.5$ cm;
$W_y = I_y/(h/2) = 2449174/53.8 = 45523\ \text{cm}^3$;
$\lambda_1 = 93.9\varepsilon = 93.9 \times 0.92 = 86.4$;
$\bar{\lambda}_{c0} = \bar{\lambda}_{LT,0} + 1.0 = 0.4 + 0.1 = 0.5$ (see Clause 6.3.2.3 of EC 3, recommended value of $\bar{\lambda}_{LT,0} = 0.4$ maximum)

then,

$\bar{\lambda}_f = k_c \times L_c/(i_{f,z} \times \lambda_1) = (0.9 \times 300)/(24.5 \times 86.4) = 0.13;$
$\bar{\lambda}_{c0} \times (M_{c,Rd})/M_{y,Ed} = 0.5 \times (W_y \times f_y/\gamma_{M1})/1304 = 0.5 \times (45523 \times 275/10^3)/1304$
$= 4.8 >> \bar{\lambda}_f\ (0.13).$

So, the condition is satisfied and the built-up welded section is adequate against lateral buckling. Therefore adopt for the beam: welded built-up section UB 1016 × 305 × 393 kg/m + 2Flg.pls 1000 mm × 30 mm.

12.2.2 Design of section of the column

Consider the section at a distance of 1.0 m below the centre line of the beam (i.e. the end of splay).

12.2.2.1 Characteristic moment

From final results already calculated for moments and forces:

Moment due to UDL M_{gk} = 4.9 kN m (permanent actions).
Moment due to machine load M_{Gk} = 88.7 kN m (permanent actions).
Moment due to vertical dynamic loading M_{Qk} = 189.1 kN m (variable actions).
Moment due to horizontal dynamic loading M_{Qkh} = 54.8 kN m (leading variable actions).

12.2.2.2 Ultimate design values

In combinations of loadings in the ultimate-limit-state design method, the partial factors shall be multiplied by the characteristic values. Referring to Table A1.2(A), "Design values of actions (EQU)(Set A)", of BS EN 1990: 2002(E), for values of γ for the ULS method:

$\sum \gamma_{G,j} \times G_{k,j} + \gamma_{Q,1} \times Q_{k,1} + \gamma_{Q,i} \times \psi_{0,i} \times Q_{k,i}$ [with accompanying variable (horizontal dynamic load)]

where the partial factors are

for combined (UDL + Machine dead load) γ_{Gj} = 1.35 (permanent actions);
for vertical dynamic loadings $\gamma_{Q,1}$ = 1.5 (leading variable action);
for horizontal dynamic loadings $\gamma_{Q,i}$ = 1.5;

and ψ = 0.7 (see Table A1.1 BS EN 1990: 2002(E)), the maximum ultimate design moment is

$$M_{Ed} = \gamma_{Gj} \times (M_{gk} + M_{Gk}) + \gamma_{Q,1} \times M_{Qk}$$
$$= 1.35 \times (4.9 + 88.7) \cdot 2 + 1.5 \times 189.1 + 1.5 \times 0.7 \times 54.8$$
$$= 467.6 \text{ kN m},$$

and the ultimate design shear at the end of splay is

$$V_{Ed} = 1.35 \times (1.8 + 31.6) + 1.5 \times 67.4 + 1.5 \times 0.7 \times 34.8 = 182.7 \text{ kN}.$$

So, the maximum ultimate thrust is

N_{Ed} = 1.35 × [shear at column base due to (UDL + weight of machine)]
+ 1.5 × [shear at column base due to dynamic vertical + 1.5 × 0.7 (horizontal dynamic loads)]
= 1.35 × (17.3 + 206.5) + 1.5 × (440.0) + 1.5 × 0.7 × 54.9 = 997.4 kN,
(see Figs. 12.3 and 12.4).

12.2.2.3 Initial sizing of section

Try the section: UB 1016 × 305 × 393 kg/m + 2Flg.pls 1000 × 30.

Properties of section

Depth of section h = 1076 mm.
Depth between fillets d = 956 mm.
Width of section b = 1000 mm.
Thickness of web t_w = 24.4 mm.
Average thickness of flange t_f = 37 mm.

Root radius r = 30 mm.
Area of section A = 1100 cm^2.
Moment of inertia I_y = 2449174 cm^4.
Moment of inertia I_z = 520500 cm^4.
Section modulus W_y = 45523 cm^3.
Section modulus W_z = 10410 cm^3.
Radius of gyration i_y = $\sqrt{(I_y/A)}$ = 47.2 cm.
Radius of gyration i_z = $\sqrt{(I_z/A)}$ = 21.8 cm.

12.2.2.4 Section classification

Flange

Assumed,

f_y = 275 N/mm^2 for steel grade S 275,
Stress factor $\varepsilon = \sqrt{(235/275)} = 0.92$ $t \leq 40$ mm,
c = 1000/2 − 303/2 = 348.5 mm,
Ratio c/t_f = 348.5/30 = 11.6,
$9\varepsilon/\alpha$ = 9 × 0.92 = 8.3.

For class 1 classification:

limiting value of $c/t_f \leq 9\varepsilon$.

In our case, c/t_f (11.6) > $9\varepsilon/\alpha$ (8.3), so the condition for class 1 section classification is not satisfied. Therefore, stiffeners shall be provided to prevent the buckling of the flange plate.

Web

Ratio $h_w/t_w = 956/24.4 = 39.2$,

and

$36 \times \varepsilon/\alpha = 36 \times 0.92/0.5 = 66.3$.

For class 1 section classification:

limiting value of $d/t_w \leq 36 \times \varepsilon/\alpha$

In our case, d/t_w (39.2) < $36 \times \varepsilon/\alpha$ (66.24). So, the condition is satisfied.

12.2.2.5 Moment capacity

Maximum ultimate design moment at the end of splay M_{Ed} = 467.6 kN m.
Maximum ultimate shear at the end of splay V_{Ed} = 182.7 kN.
Maximum ultimate thrust at the end of splay N_{Ed} = 997.4 kN.

We have already seen in the section design of the beam that the above value of the ultimate design moment M_{Ed} = 467.6 kN m is much lower than the moment capacity of the section. So, it is not necessary to verify the section.

12.2.2.6 Buckling resistance of the member subjected to uniform compression

Referring to Clause 6.3.1.1, the compression member shall be verified against buckling by the following expression:

$$N_{Ed}/N_{b,Rd} \leq 1.0 \quad (6.46)$$

where

N_{Ed} = ultimate design compression,
$N_{b,Rd}$ = ultimate buckling resistance of the compression member.

Now, the design buckling resistance may be expressed by the following equation:

$$N_{b,Rd} = \chi \times A \times f_y/\gamma_{M1} \quad (6.47)$$

where

χ = reduction factor for the relevant buckling mode
$= 1/[\Phi + \sqrt{(\Phi^2 - \bar{\lambda}^2)}]$ (6.49)

and

$\Phi = 0.5[1 + \alpha(\bar{\lambda} - 2) + \bar{\lambda}^2]$,
α = imperfection factor (value may be obtained from Table 6.1 and Table 6.2 of EC 3, Part 1-1 – see Appendix),

$$\bar{\lambda} = L_{cr}/(i_z \times \lambda_1). \quad (6.50)$$

Using

L_{cr} = buckling length in the buckling plane considered $= 0.75H = 0.75 \times 5.71$
= 430 cm (the end is assumed fixed in both axes; H = height of column),
i_z = radius of gyration about the minor axis = 21.8 cm,
$\lambda_1 = 93.9\varepsilon = 93.9 \times 0.92 = 86.4$,

we determine

$$\bar{\lambda} = 430/(21.8 \times 86.4) = 0.23.$$

With a welded built-up section $t_f > 40$ and the $z-z$ axis, follow curve "d" in Fig. 6.4 of EC 3, Part 1-1. Referring to Table 6.1 (see Appendix), with buckling curve "d", the imperfection factor $\alpha = 0.76$. So,

$$\Phi = 0.5[1 + 0.76 \times (0.23 - 0.2) + 0.23^2] = 0.54.$$

Thus

$$\chi = 1/[0.54 + \sqrt{(0.54^2 - 0.23^2)}] = 1/1.03 = 0.97,$$

and

$$N_{b,Rd} = \chi \times A \times f_y/\gamma_{M1} = 0.97 \times 1100 \times 100 \times 275/10^3$$
$$= 29343 \text{ kN} >> N_{Ed} \text{ (997.4 kN)}.$$ Satisfactory

Therefore, adopt the column section: UB 1016 × 305 × 393 kg/m + 2Flg.pls 1000 mm × 30 mm.

12.3 Design of weld

In the design of weld, EC 3, Part 1-8, "Design of joints", will be followed (Eurocode, 2005). All the references and tables in what follows refer to this Eurocode, except where mentioned otherwise.

12.3.1 Connection between the flange plate and the UB section

12.3.1.1 Fillet weld

The horizontal shear/linear length = vertical shear/linear depth. So, we shall calculate the maximum horizontal shear stress due to maximum vertical ultimate shear by the approximate formula:

Vertical shear/linear height of girder = V_{Ed}/h = 1008/1016 = 0.99 kN.
Therefore, the design horizontal shear/linear length of girder $F_{W,Ed}$ = 0.99 kN.

Referring to Clause 4.5.3.3 ("Simplified method for design resistance of fillet weld"), the design shear strength is expressed as:

$$f_{vw,d} = f_u/\sqrt{3}/(\beta_w \times \gamma_{M2}). \tag{4.4}$$

Referring to Table 3.1, for grade S 275 steel,

f_u = 430 N/mm^2,

and referring to Table 4.1, for grade S 275 steel,

β_w = 0.85.

Referring to Clause 6.1, the partial factor for resistance to joints γ_{M2} = 1.25. Therefore

$f_{vw,d}$ = 430/√3/(0.85 × 1.25) = 234 N/mm^2.

Assuming the size of the fillet weld s = 12 mm and the thickness of the throat a = 0.7 × 12 = 8.4 mm, the weld resistance/mm length is therefore

$F_{W,Rd} = f_{vw,d} \times a \times 2 = 234 \times 8.4 \times 2/10^3$
= 3.9 kN/mm >> $F_{W,Ed}$ (0.99 kN). Satisfactory

Therefore, adopt a 12 mm fillet weld in the connection between the flange plates and the girder. For details of frame 2 see Fig. 12.5.

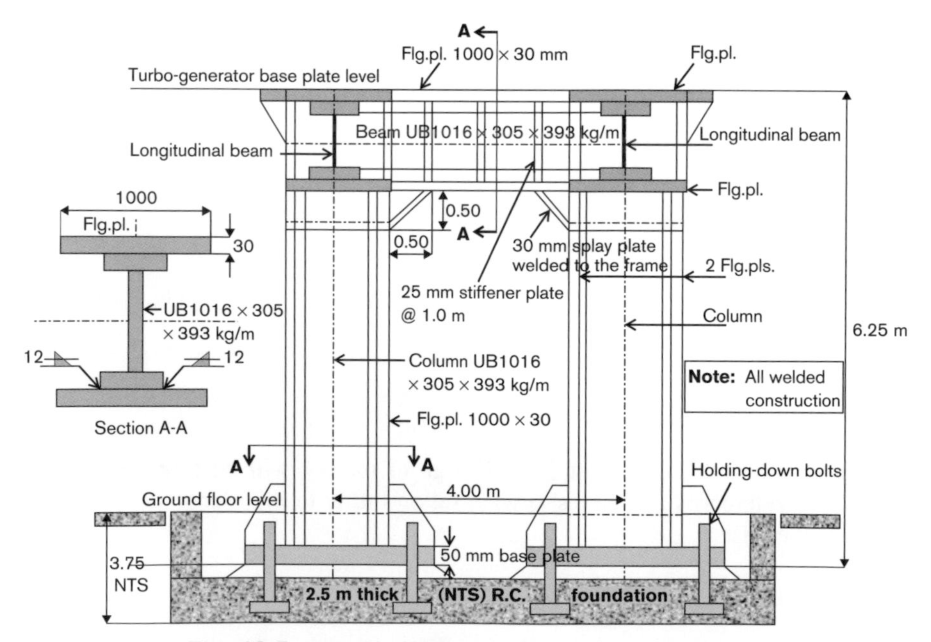

Fig. 12.5. Details of turbo-generator support frame 2

References

Eurocode, 2002. BS EN 1990: 2002(E), Eurocode Basis of Structural Design.

Eurocode, 2005. BS EN 1993-1-1: 2005, Eurocode 3. Design of Steel Structures.

Eurocode, 2005. BS EN 1993-1-8: 2005, Eurocode 3. Design of Steel Structures. Design of Joints.

Kleinlogel, A., 1958. *Rahmenformeln "Rigid Frame Formulae"*. Wilhelm Ernst & Sohn, Berlin.

APPENDIX

Annex A of Eurocode 3, Part 1-1, BS EN 1993-1-1: 2005

In this appendix, we reproduce Tables 3.1, 5.2, 6.1, 6.2, 6.3 and 6.4 and Fig. 6.4 contained in Annex A of Eurocode 3, Part 1-1, and also Tables A1.1 and A1.2(B) of BS EN 1990: 2002(E).

EN 1993-1-1: 2005 (E)

(3)B For building components under compression a minimum toughness property should be selected.

NOTE B The National Annex may give information on the selection of toughness properties for members in compression. The use of Table 2.1 of EN 1993-1-10 for σ_{Ed} = 0,25 $f_y(t)$ is recommended.

(4) For selecting steels for members with hot dip galvanized coatings see EN 1461.

Table 3.1: Nominal values of yield strength f_y and ultimate tensile strength f_u for hot rolled structural steel

Standard and steel grade	Nominal thickness of the element t [mm]			
	$t \leq 40$ mm		40 mm $< t \leq 80$ mm	
	f_y [N/mm²]	f_u [N/mm²]	f_y [N/mm²]	f_u [N/mm²]
EN 10025-2				
S 235	235	360	215	360
S 275	275	430	255	410
S 355	355	510	335	470
S 450	440	550	410	550
EN 10025-3				
S 275 N/NL	275	390	255	370
S 355 N/NL	355	490	335	470
S 420 N/NL	420	520	390	520
S 460 N/NL	460	540	430	540
EN 10025-4				
S 275 M/ML	275	370	255	360
S 355 M/ML	355	470	335	450
S 420 M/ML	420	520	390	500
S 460 M/ML	460	540	430	530
EN 10025-5				
S 235 W	235	360	215	340
S 355 W	355	510	335	490
EN 10025-6				
S 460 Q/QL/QL1	460	570	440	550

Table 5.2 (sheet 1 of 3). Maximum width-to-thickness ratios for compression parts [from EC 3, Part 1-1, BS EN 1993-1-1: 2005]

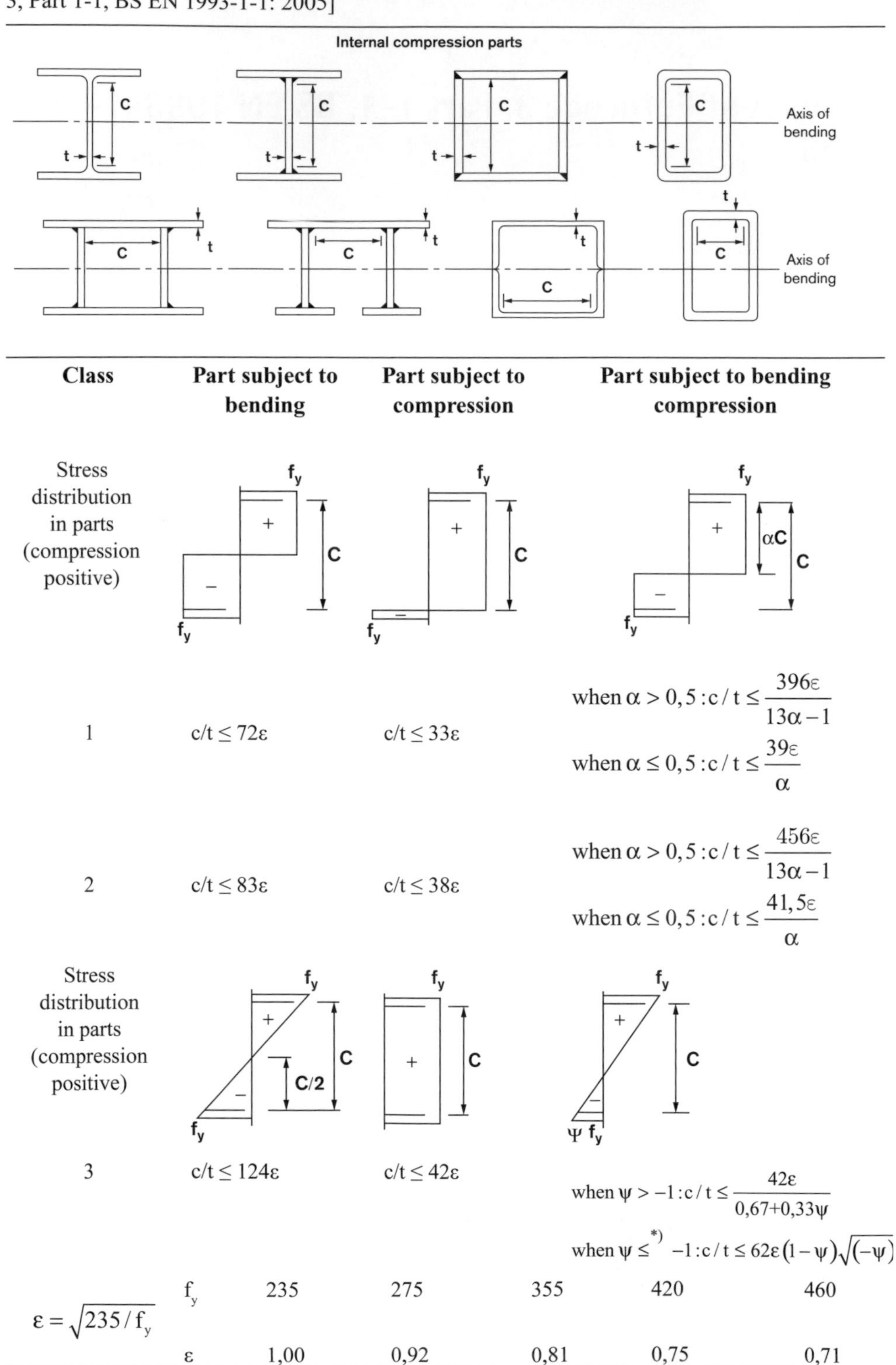

Class	Part subject to bending	Part subject to compression	Part subject to bending compression
Stress distribution in parts (compression positive)			
1	$c/t \le 72\varepsilon$	$c/t \le 33\varepsilon$	when $\alpha > 0,5 : c/t \le \dfrac{396\varepsilon}{13\alpha - 1}$ when $\alpha \le 0,5 : c/t \le \dfrac{39\varepsilon}{\alpha}$
2	$c/t \le 83\varepsilon$	$c/t \le 38\varepsilon$	when $\alpha > 0,5 : c/t \le \dfrac{456\varepsilon}{13\alpha - 1}$ when $\alpha \le 0,5 : c/t \le \dfrac{41,5\varepsilon}{\alpha}$
Stress distribution in parts (compression positive)			
3	$c/t \le 124\varepsilon$	$c/t \le 42\varepsilon$	when $\psi > -1 : c/t \le \dfrac{42\varepsilon}{0,67+0,33\psi}$ when $\psi \le^{*)} -1 : c/t \le 62\varepsilon(1-\psi)\sqrt{(-\psi)}$

$\varepsilon = \sqrt{235/f_y}$	f_y	235	275	355	420	460
	ε	1,00	0,92	0,81	0,75	0,71

*) $\psi \le -1$ applies where either the compression stress $\sigma \le f_U$ or the tensile strain $\varepsilon_U > f_U/E$

Table 5.2 (sheet 2 of 3). Maximum width-to-thickness ratios for compression parts [from EC 3, Part 1-1, BS EN 1993-1-1: 2005]

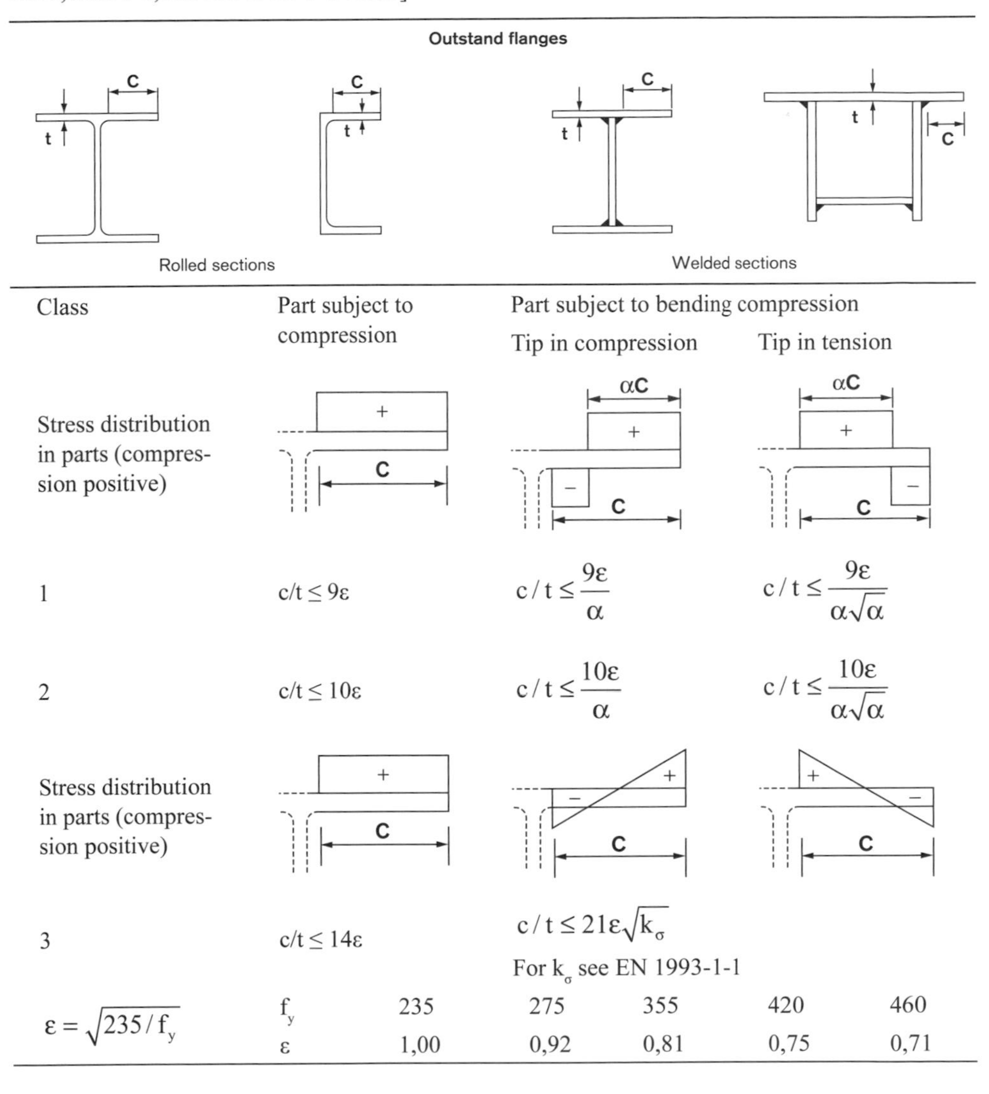

Class	Part subject to compression	Part subject to bending compression: Tip in compression	Part subject to bending compression: Tip in tension
Stress distribution in parts (compression positive)			
1	$c/t \leq 9\varepsilon$	$c/t \leq \frac{9\varepsilon}{\alpha}$	$c/t \leq \frac{9\varepsilon}{\alpha\sqrt{\alpha}}$
2	$c/t \leq 10\varepsilon$	$c/t \leq \frac{10\varepsilon}{\alpha}$	$c/t \leq \frac{10\varepsilon}{\alpha\sqrt{\alpha}}$
Stress distribution in parts (compression positive)			
3	$c/t \leq 14\varepsilon$	$c/t \leq 21\varepsilon\sqrt{k_\sigma}$ For k_σ see EN 1993-1-1	

$\varepsilon = \sqrt{235/f_y}$						
	f_y	235	275	355	420	460
	ε	1,00	0,92	0,81	0,75	0,71

Table 5.2. (sheet 3 of 3). Maximum width-to-thickness ratios for compression parts [from EC 3, Part 1-1, BS EN 1993-1-1: 2005]

	Angles	
Refer also to "Outstand flanges" (see sheet 2 of 3)	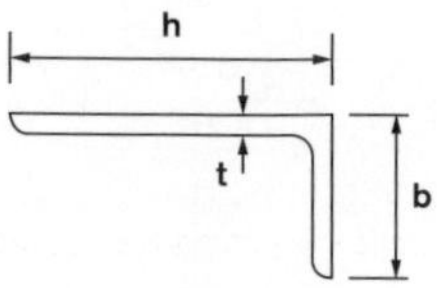 h, t, b	Does not apply to angles in continuous contact with other components
Class	Section in compression	
Stress distribution across section (compression positive)	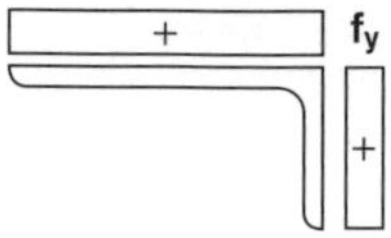 +, f_y, +	
3	$h/t \leq 15\varepsilon : \frac{b+h}{2t} \leq 11{,}5\varepsilon$	

Tubular sections

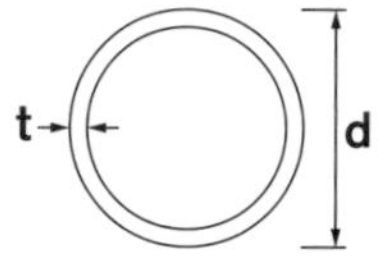

t, d

Class	Section in bending and/or compression
1	$d/t \leq 50\varepsilon^2$
2	$d/t \leq 70\varepsilon^2$
3	$d/t \leq 90\varepsilon^2$ **NOTE** For $d/t > 90\varepsilon^2$ see EN 1993-1-6.

$\varepsilon = \sqrt{235/f_y}$	f_y	235	275	355	420	460
	ε	1,00	0,92	0,81	0,75	0,71
	ε^2	1,00	0,85	0,66	0,56	0,51

Table 6.1. Imperfection factors for buckling curves [from Eurocode 3, Part 1-1, BS EN 1993-1-1: 2005]

Buckling curve	a_g	a	b	c	d
Imperfection factor α	0.13	0.21	0.34	0.49	0.76

Table 6.2: Selection of buckling curve for a cross-section [from Eurocode 3, Part 1-1, BS EN 1993-1-1: 2005]

Cross section		Limits	Buckling about axis	Buckling Curve: S 235, S 275, S 355, S 420	Buckling Curve: S 460
Rolled sections	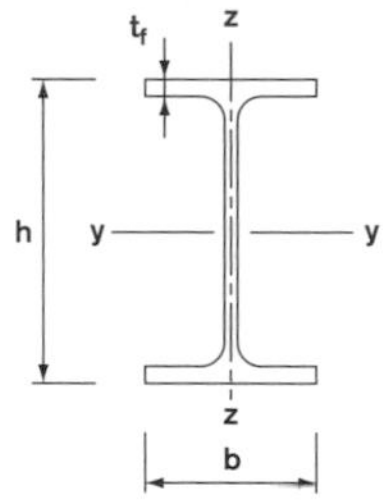 $h/b > 1,2$	$t_f \leq 40$ mm	y-y	a	a_0
			z-z	b	a_0
		40 mm $< t_f \leq 100$	y-y	b	a
			z-z	c	a
	$h/b \leq 1,2$	$t_f \leq 100$ mm	y-y	b	a
			z-z	c	a
		$t_f > 100$ mm	y-y	d	c
			z-z	d	c
Welded I-sections	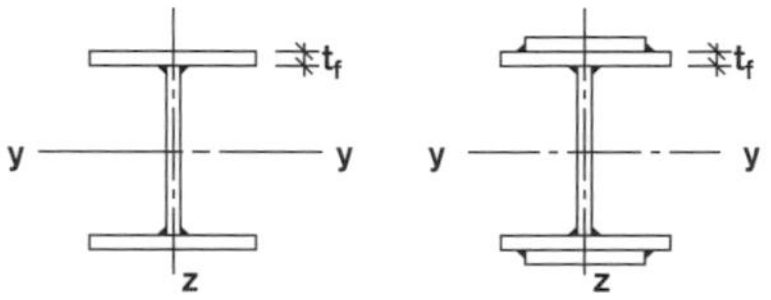	$t_f \leq 40$ mm	y-y	b	b
			z-z	c	c
		$t_f > 40$ mm	y-y	c	c
			z-z	d	d
Hollow sections	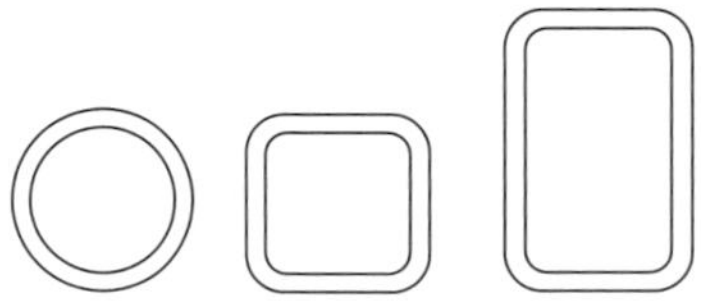	hot finished	any	a	a_0
		cold formed	any	c	c
Welded box sections	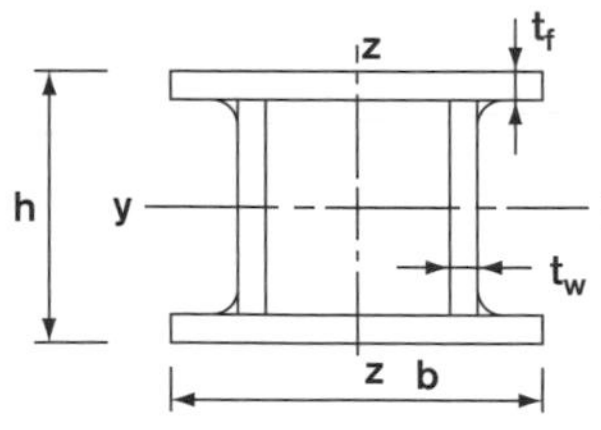	generally(except as below)	any	b	b
		thick welds: $a > 0,5t_f$; $b/t_f < 30$; $h/t_w < 30$	any	c	c
U-, T- and solid sections			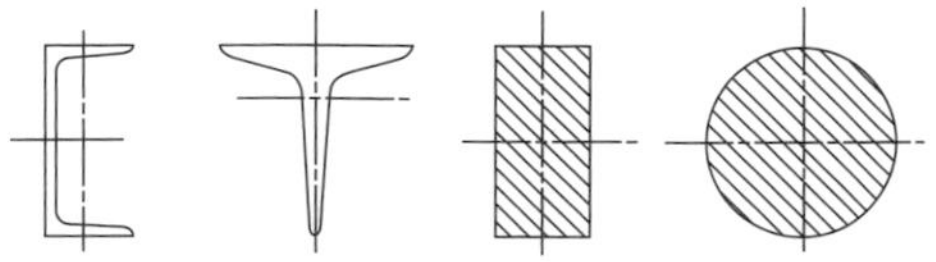 any	c	c
L- sections			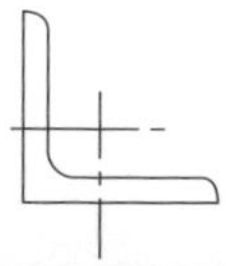 any	b	b

Table 6.3. Recommended values for imperfection factors for lateral torsional buckling curves [from Eurocode 3, Part 1-1, BS EN 1993-1-1: 2005]

Buckling curve	a	b	c	d
Imperfection factor α_{LT}	0.21	0.34	0.49	0.76

Note: The recommendations for buckling curves in Fig. 6.4 are given in Table 6.4.

Table 6.4. Recommended values for lateral torsional buckling curves for cross-sections using equation (6.56) [from Eurocode 3, Part 1-1, BS EN 1993-1-1: 2005]

Cross-section	Limits	Buckling curve
Rolled I-sections	$h/b \leq 2$	**a**
	$h/b > 2$	**b**
Welded I-sections	$h/b \leq 2$	**c**
	$h/b > 2$	**d**
Other cross-sections	-	**d**

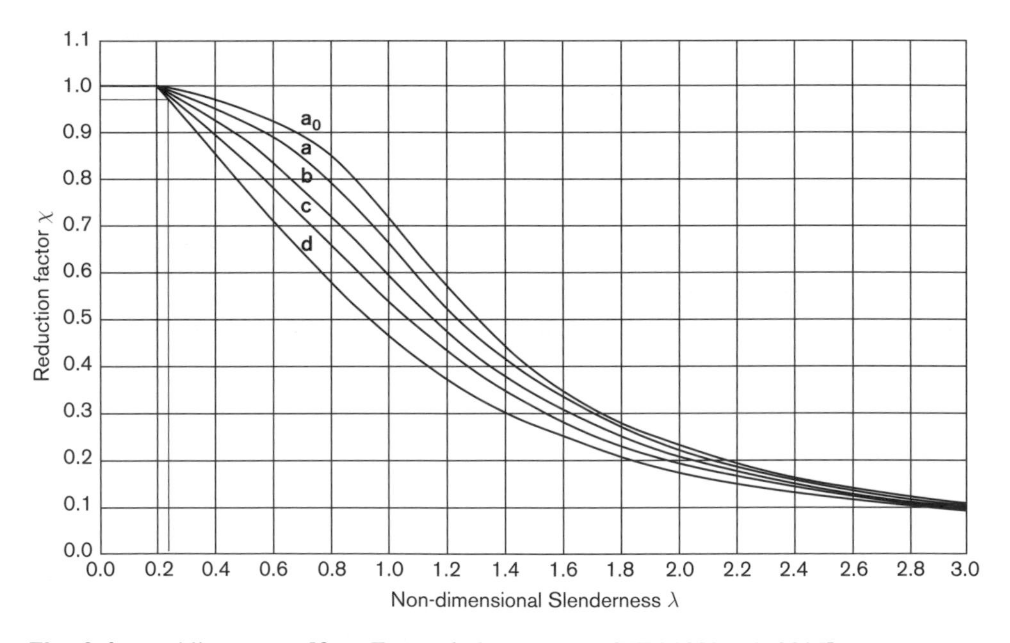

Fig. 6.4. Buckling curves [from Eurocode 3, Part 1-1, BS EN 1993-1-1: 2005]

Note: The imperfection factor corresponding to the appropriate buckling curve in Fig. 6.4 should be obtained from Table 6.1 and Table 6.2.

Table A1.1. Recommended values of ψ factors for buildings [from BS EN 1990: 2002(E)]

Action	ψ_0	ψ_1	ψ_2
Imposed loads in buildings, category (see EN 1991-1-1)			
Category A: domestic, residential areas	0,7	0,5	0,3
Category B: office areas	0,7	0,5	0,3
Category C: congregation areas	0,7	0,7	0,6
Category D: shopping areas	0,7	0,7	0,6
Category E: storage areas	1,0	0,9	0,8
Category F: traffic area, vehicle weight ≤ 30kN	0,7	0,7	0,6
Category G: traffic area, 30kN < vehicle weight ≤ 160kN	0,7	0,5	0,3
Category H: roofs	0	0	0
Snow loads on buildings (see EN 1991-1-3)*			
Finland, Iceland, Norway, Sweden	0,70	0,50	0,20
Remainder of CEN Member States, for sites located at altitude H > 1000 m a.s.l.	0,70	0,50	0,20
Remainder of CEN Member States, for sites located at altitude H ≤ 1000 m a.s.l.	0,50	0,20	0
Wind loads on buildings(see EN 1991-1-4)	0,6	0,2	0
Temperature (non-fire) in buildings (see EN 1991-1-5)	0,6	0,5	0
NOTE The ψ values may be set by the National annex. *For countries not mentioned below, see relevant local conditions.			

Table A1.2(B). Design values of actions (STR/GEO) (Set B) [from BS EN 1990: 2002(E)]

Persistent and transient design situations	Permanent actions		Loading variable action	Accompanying variable actions*	
	Unfavourable	*Favourable*		*Main (if any)*	*Others*
Equation 6.10	$\gamma_{Gj,sup}G_{kj,sup}$	$\gamma_{Gj,inf}G_{kj,inf}$	$\gamma_{Q,1}Q_{k,1}$	-	$\gamma_{Q,1}\psi_{0,1}Q_{k,i}$
Equation 6.10a	$\gamma_{Gj,sup}G_{kj,sup}$	$\gamma_{Gj,inf}G_{kj,inf}$	-	$\gamma_{Q,1}\psi_{0,1}Q_{k,1}$	$\gamma_{Q,1}\psi_{0,1}Q_{k,i}$
Equation 6.10b	$\zeta\gamma_{Gj,sup}G_{kj,sup}$	$\gamma_{Gj,inf}G_{kj,inf}$	$\gamma_{Q,1}Q_{k,1}$	-	$\gamma_{Q,1}\psi_{0,1}Q_{k,i}$

* Variable actions are those considered in Table A1.1

The following values for γ and ζ are recommended when using expressions 6.10, 6.10a and 6.10b:

$\gamma_{Gj,sup} = 1.35$;
$\gamma_{Gj,inf} = 1.0$;
$\gamma_{Q,1} = 1.5$ where unfavourable (0 where favourable);
$\gamma_{Q,i} = 1.5$ where unfavourable (0 where favourable);
$\zeta = 0.85$ (so that $\zeta\gamma_{Gj,sup} = 0.85 \times 1.35 = 1.15$).

Further Reading

Books

American Institute of Steel Construction, 2006. *Manual of Steel Construction*, 13th edn., American Institute of Steel Construction Inc, New York, USA.

Gaylord, Jr. E.H. and Gaylord, N. Charles, 1979. *Structural Engineering Handbook*, McGraw-Hill Book Company Inc, New York.

Ghosh, Karuna Moy, 2010. *Practical Design of Steel Structures*, Whittles Publishing, Caithness.

Ghosh, K.M., 2010. *Practical Design of Reinforced Concrete Structures*, Prentice Hall Learning Private Limited, New Delhi, India.

Major, Alexander, 1962. *Vibration Analysis and Design of Foundations for Machines and Turbines*, Collet's Holdings Limited, London.

Murphy, Glenn, 1948. *Properties of Engineering Materials*, International Textbook Company, Pennsylvania, USA.

Pippard, A.J.S. and Baker, J.F., 1953. *The Analysis of Engineering Structures*, Edward Arnold & Co., London.

Richart, Jr. F.E., Hall, Jr. J.R. and Woods, R.D., 1969. *Vibrations of Soils and Foundations*, Prentice-Hall Inc., N.J, USA.

Steel Construction Institute, 2003. *Steel Designers' Manual*, 6th edn, Blackwell Science, Oxford, UK.

Wiegel, R.L., 1970. *Earthquake Engineering*, Prentice-Hall Inc., N.J, USA.

Winterkorn, F. Hans and Fang, Hsai-Yang, 1970. *Foundation Engineering Handbook*, Van Nostrand Rheinhold Company, New York.

Papers:

Elghazouli, A., 2007. Seismic Design of Steel Structures to EC8; *The Structural Engineer*, 85(6).

Ingham, Jason and Griffith, M., 2011. Damage to Unreinforced Masonry Structures by Seismic Activity, *The Structural Engineer*, 89(3).

Macabuag, Josha, 2009. Seismic Reinforcement of Adobe in Rural Peru. *The Structural Engineer*, 88(12).

Mann, Allan, Quinion, David, Bolton, Chris and Powderham, Alan, 2007. Nuclear Power Station Design, *The Structural Engineer* 85(7).

News, Journal, The Structural Engineer, 2011. New Zealand City of Christchurch Hit by 6.3 Magnitude Earthquake, *The Structural Engineer*, 89(3).

Prasad, A.M and Menon, D., 2008. Earthquake Resistant Design Practice and Research in India, *The Structural Engineer*, 86(9).

Index